SPSS for Windows Manual
for

David S. Moore's

The Basic Practice of Statistics

Third Edition

LINDA SORENSEN
Algoma University College

W. H. Freeman and Company
New York

ISBN 0-7167-5884-9

Printed in the United States of America

First printing 2003

Contents

Preface

With the use of software almost mandatory today in the teaching and application of statistics, the introductory statistics course has changed from an emphasis on algebra and mathematics to concentrating on logic. This is a marvelous leap forward. It makes learning statistics quite different from my undergraduate days, and if, like me, you are not a good mathematician, it makes it much more interesting. I learned to love logic and representations for things by watching my father build radios from schematic diagrams when I was quite young. It amazed me that a thing that looked like circles and squares on a page could represent the working parts of a radio. It was also my first understanding that the parts had to go together in a logical way or they wouldn't work. Perhaps this is why statistics today, with all the advantages of software packages, is so intriguing to me.

This manual is a supplement to the third edition of Moore's *The Basic Practice of Statistics* (BPS). The purpose of this manual is to show students how to perform the statistical procedures discussed in BPS using SPSS 11.0 for Windows. This manual provides applications and examples for each chapter of the text. The process was guided by the principles put forth by the American Statistical Association for teaching statistics. You will observe that (for consistency with BPS) most of the examples and subsequent discussion come directly from BPS. Step-by-step instructions describing how to carry out statistical analyses using SPSS 11.0 for Windows are provided.

SPSS has been regarded as one of the most powerful statistical packages for many years. It performs a wide variety of statistical techniques ranging from descriptive statistics to complex multivariate procedures. In addition, a number of improvements have been made to version 11.0 of SPSS for Windows that make it more user-friendly.

Writing this manual has been a learning experience for me, and in the words of Canadian author Frank Paci, good teachers are good learners. I would like to think that I've been a good learner and that this will do two things: I believe that it will improve my teaching, and I believe that it will improve your learning.

I am indebted to the authors of earlier editions of this manual for their thoroughness and attention to detail. I would like to thank Danielle Swearengin and Christopher Spavins of W. H. Freeman and Company, for giving me the opportunity to undertake this project. Christopher talked me into it, and Danielle talked me through it. While I take full responsibility for errors and omissions, I truly appreciate the indefatigable assistance of Sarah Campbell. Sarah helped me get through it.

I dedicate this work to my father, Ross Hunter (1926–2003).

Chapter 0. Introduction to SPSS 11.0 for Windows

Topics covered in this chapter:

- **Accessing SPSS 11.0 for Windows**
- **Entering Data**
- **Saving an SPSS for Windows Data File**
- **Opening an Existing SPSS for Windows Data File from Disk**
- **Opening an SPSS Data File from the CD-ROM Accompanying BPS**
- **Opening a Microsoft Excel Data File from Disk**
- **Defining a Variable**
- **Recoding a Variable**
- **Deleting a Case from an SPSS for Windows Data File**
- **Opening Data Sets Not Created by SPSS or Windows Excel**
- **Printing in SPSS for Windows**
- **Copying from SPSS for Windows into Microsoft Word 97**
- **Using SPSS for Windows Help**

This manual is a supplement to *The Basic Practice of Statistics*, Third Edition, by David S. Moore, which is referred to as BPS throughout the manual. The purpose of this manual is to show students how to perform the statistical procedures discussed in BPS using SPSS 11.0 for Windows. This manual is not meant to be a comprehensive guide to all procedures available in SPSS for Windows. The instructions included here will work for most versions of SPSS for Windows.

Throughout this manual, the following conventions are used: (1) variable names are given in boldface italics (e.g., ***age***); (2) commands you click or text you type are boldface (e.g., click **Analyze**); (3) important statistical terms are boldface; (4) the names of boxes or areas within an SPSS for Windows window are in double quotes (e.g., the "Variable Name" box); and (5) in an example number, the digit(s) before the decimal place is the chapter number, the digit(s) after the decimal place is the example number within that chapter (e.g., Example 1.3 is the third example in Chapter 1). Unless otherwise specified, all example, table, and figure numbers refer to examples, tables, and figures within this manual.

This chapter serves as a brief overview of tasks such as entering data, reading in data, saving data, printing output, and using SPSS for Windows Help that will help you get started in SPSS for Windows.

Accessing SPSS 11.0 for Windows

Find out how to access SPSS for Windows at your location. Figure 0.1 shows the opening screen in SPSS 11.0 for Windows after the software has been activated. Note that in the section labeled "Open an existing file" located within the "SPSS for Windows" window, the most recent files used in SPSS for Windows are listed, and you can open them easily by clicking on the desired file name and then clicking " OK." If you want to enter data into a new file, you can click on "Type in data" and then click " OK."

For new SPSS for Windows users, I recommend clicking on "Run the tutorial" to become acquainted with the software package.

The window in the background of Figure 0.1 is the SPSS for Windows Data Editor. You will note that it is in a spreadsheet format, where columns represent variables and rows represent individual cases or observations. The SPSS for Windows Menu bar (**File**, **Edit**, ... , **Help**) appears directly below "Untitled – SPSS for Windows Data Editor." Each of these main menu options contains its own submenu of additional options. Throughout this manual, the first step in performing a particular task or analysis typically gives the SPSS for Windows Menu bar as the starting point (e.g., Click **File**, then click **Open**).

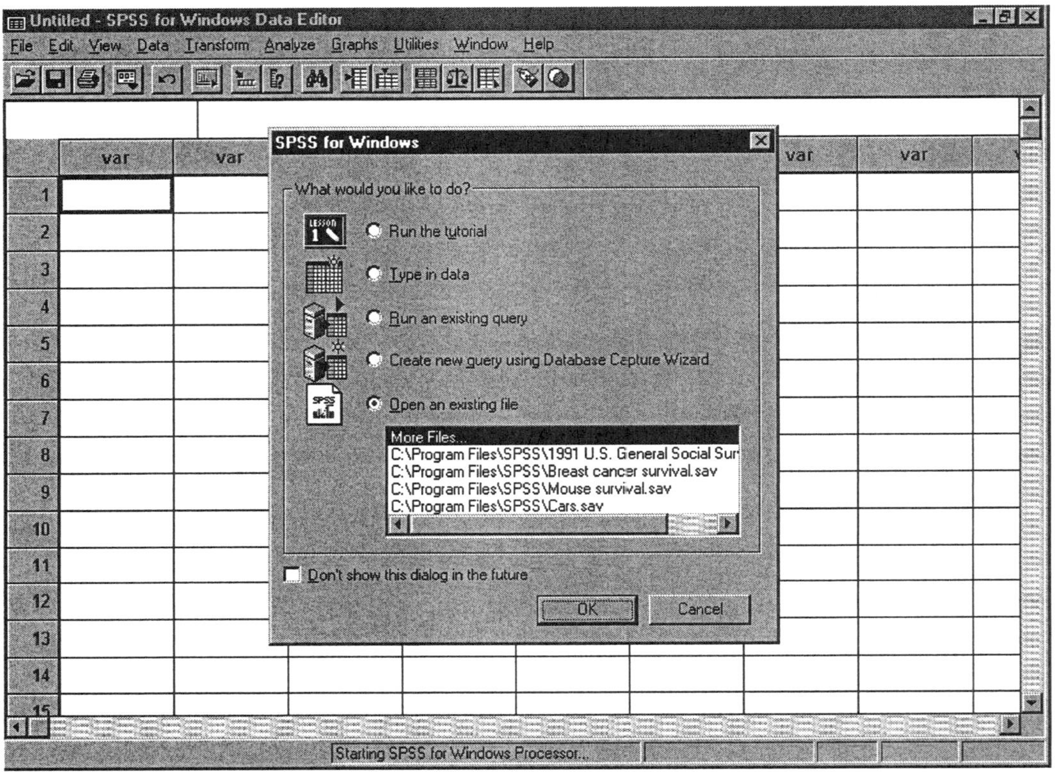

Figure 0.1

In addition to the Data Editor, the other primary window is the Output – SPSS for Windows Viewer window (which is not accessible until output has been generated). To move between these two windows, select **Window** from the SPSS for Windows Main Menu and click on the name of the desired window. In Figure 0.2, the SPSS for Windows Data Editor is the current active window, as shown by the ✔ in front of that window name.

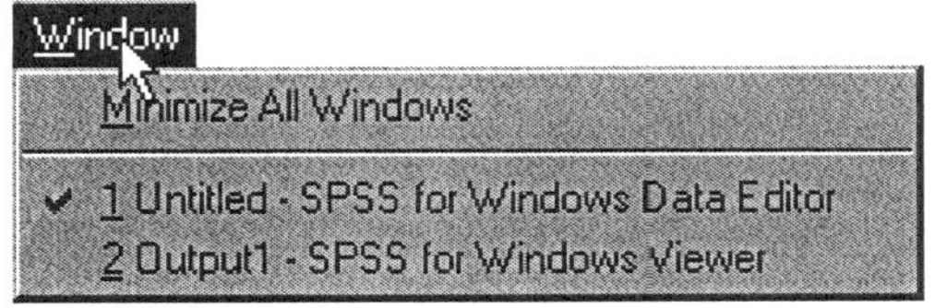

Figure 0.2

Most of the output in SPSS for Windows can be generated by clicking on a series of commands using a sequence of pulldown menus. However, it is also possible to generate output by writing a program using

SPSS for Windows syntax language. The explanation of the syntax language is beyond the scope of this manual.

Entering Data

Before entering a data set into SPSS for Windows, determine whether the variable is quantitative or categorical. A **categorical** variable places an individual into one of several groups or categories, and a **quantitative** variable assumes numerical values for which arithmetic operations make sense. SPSS for Windows further classifies variables as ordinal, nominal, or scale. An **ordinal** variable classifies characteristics about the objects under study into categories that can be logically ordered. Some examples of ordinal variables are the size of an egg (small, medium, or large) and class standing (freshman, sophomore, junior, or senior). A **nominal** variable classifies characteristics about the objects under study into categories. Some examples of nominal variables are eye color, race, and gender. **Scale** variables collectively refer to both interval and ratio variables and are quantitative variables for which arithmetic operations make sense. Some examples of scale variables are height, weight, and age. By default, SPSS for Windows assumes that any new variable is a scale variable formatted to have a width of 8 digits, 2 of which are decimal places. This is denoted in SPSS for Windows by "numeric 8.2."

Suppose you want to enter the data set presented in Example 1.2 in BPS into SPSS for Windows. The variables in this data set include one categorical and two quantitative variables.

To create this data set in SPSS for Windows, follow these steps:

1. Click **File,** click **New,** and then click **Data.** The SPSS for Windows Data Editor is cleared.
2. Enter the first few rows of data from Table 1.2 (see Figure 0.3).

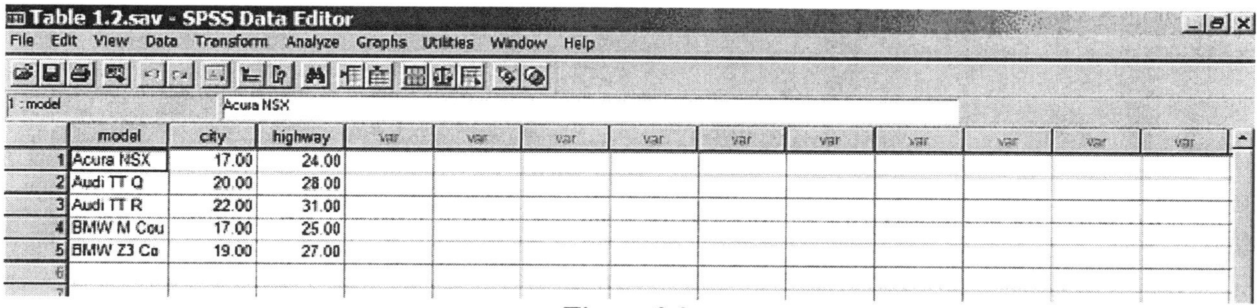

Table 1.2.sav - SPSS Data Editor

File Edit View Data Transform Analyze Graphs Utilities Window Help

1 : model Acura NSX

	model	city	highway	var	var	var	var	var	var	var	var	var	var
1	Acura NSX	17.00	24.00										
2	Audi TT Q	20.00	28.00										
3	Audi TT R	22.00	31.00										
4	BMW M Cou	17.00	25.00										
5	BMW Z3 Co	19.00	27.00										
6													
7													

Figure 0.3

3. To define the variables, click on the **Variable View** tab at the bottom of the screen (See Figure 0.4) and make the necessary adjustments.

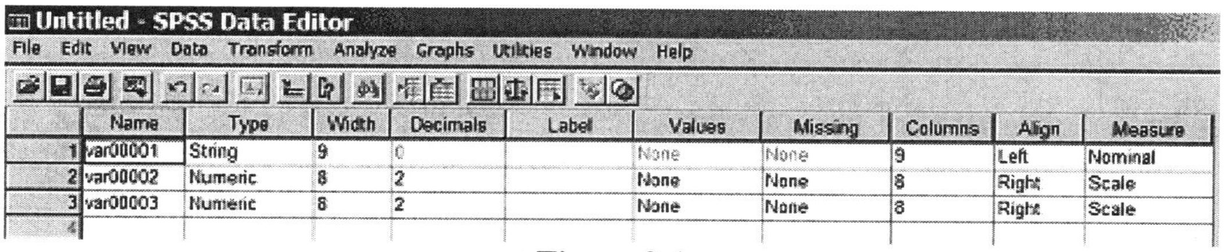

Untitled - SPSS Data Editor

File Edit View Data Transform Analyze Graphs Utilities Window Help

	Name	Type	Width	Decimals	Label	Values	Missing	Columns	Align	Measure
1	var00001	String	9	0		None	None	9	Left	Nominal
2	var00002	Numeric	8	2		None	None	8	Right	Scale
3	var00003	Numeric	8	2		None	None	8	Right	Scale
4										

Figure 0.4

4. To change the variable name from the default name of ***var00001*** to a more appropriate name (such as ***model***), type ***model*** into the "Name" box. Variable names in SPSS for Windows can be at most 8 characters long, containing no embedded blanks.
5. To change the type of the variable, click in the **Type** box and then on the three dots that appear. The "Variable Type" window shown in Figure 0.5 appears.

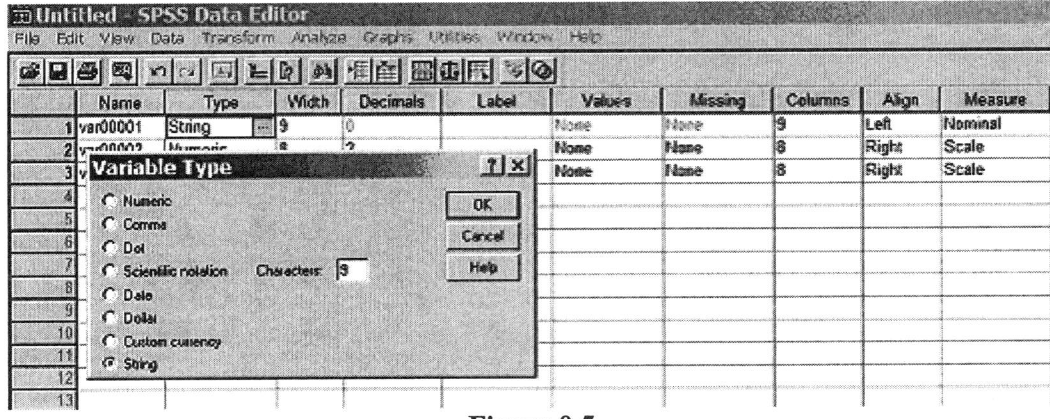

Figure 0.5

6. Click **String.**
7. To change the width of the variable ***model*** from the default width of 8 characters to 9 characters, type **9** in the "Width" box.
8. Click **Continue.**
9. Click **OK.** The variable name ***model*** now appears in the SPSS for Windows Data Editor.
10. Each of the other variable characteristics can also be modified as appropriate. For example, you may wish to remove the decimal places that SPSS automatically added. Click in the "Decimal" box and use the down arrow to set the number of decimal places to 0 (see Figure 0.6).

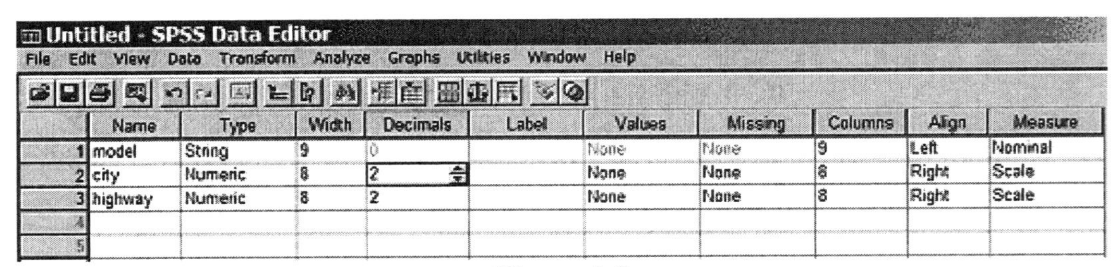

Figure 0.6

Saving an SPSS for Windows Data File

To save an SPSS for Windows data file, follow these steps:

1. Click **File** and click **Save as.** The "Save Data As" window appears (see Figure 0.7 on the next page).
2. If you wish to save the file on disk, click ▼ in the "Save in" box until **3½ Floppy [A:]** appears, and then click on this name. If you prefer to save the file in a different location, continue to click on ▼ until the name of the desired location is found, then click on this location name.
3. By default, SPSS for Windows assigns the .sav extension to data files. If you wish to save the data in a format other than an SPSS for Windows data file, click ▼ in the "Save as type" box until the name of the desired file type appears, then click on this file type name.

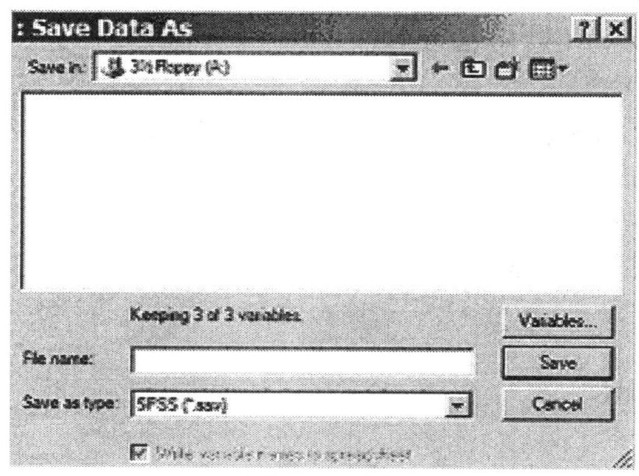

Figure 0.7

4. In the "File name" box, type in the desired name of the file you are saving. The name **Table 1.2** is used in Figure 0.8. Make sure a disk is in the A: drive, and then click **Save.**

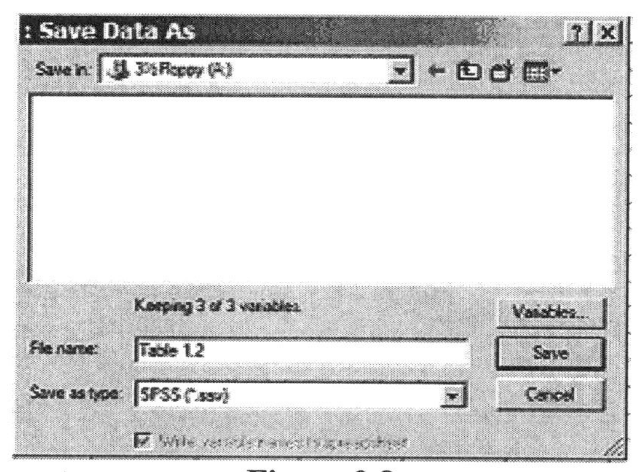

Figure 0.8

Opening an Existing SPSS for Windows Data File from Disk

To open an existing SPSS for Windows data file from a disk, follow these steps:

1. Click **File** and click **Open.** The "Open File" window appears (see Figure 0.9).
2. If you wish to open an SPSS for Windows data file from a disk in the A: drive, click ▼ in the "Look in" box until **3½ Floppy [A:]** appears, then click on this name. If you prefer to open a file stored in a different location, continue to click on ▼ until the name of the desired location is found, then click on this location name.
3. All files with a .sav extension will be listed in the window. Often you will need to click on "All files:" so that SPSS portable (.por) files appear in the window as well. Click on the name of the data file you wish to open. This name now appears in the "File name" box. In Figure 0.9, **Table 1.2.sav** is the desired SPSS for Windows data file.
4. Click **Open.**

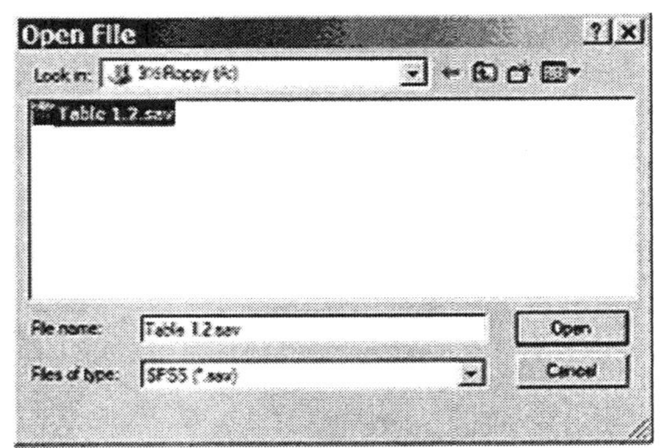

Figure 0.9

Opening an SPSS Data File from the CD-ROM Accompanying BPS

To open an SPSS Data File from the CD-ROM accompanying BPS, follow these steps:

1. From the SPSS for Windows Data Editor Menu bar, click **File** and then click **Open.** The SPSS for Windows "Open File" window appears (see Figure 0.10).

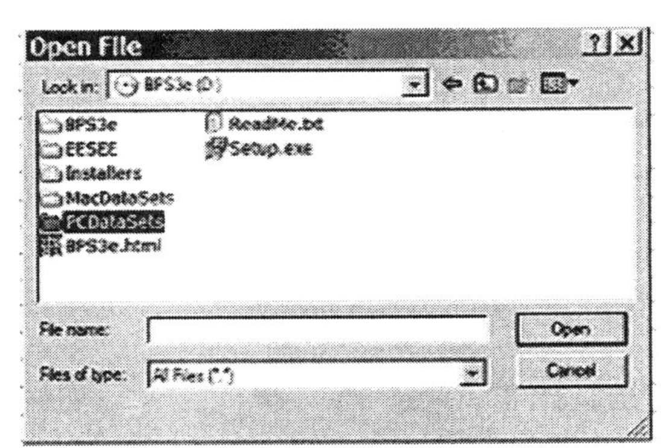

Figure 0.10

2. Click ▾ in the "Look in" box until the name **BPS3e (D:)** appears. The letter **(D:)** represents the label for the CD-ROM drive. Yours may have a different letter.
3. Next double-click on **PCDataSets,** and then double-click on **SPSS.**
4. Make sure that "All Files (*.*)" appears in the "Files of type:" box.
5. Click on the chapter number (**chap01,** for example) and then the name of the file you wish to open in SPSS for Windows. The desired file name will appear in the "File name" box. In this example the name of the desired Microsoft SPSS file is **ta01-06.por.**

Opening a Microsoft Excel Data File from Disk

To open a Microsoft Excel data file in SPSS for Windows, follow these steps:

1. From the SPSS for Windows Data Editor Menu bar, click **File** and then click **Open.** The SPSS for Windows "Open File" window appears as it did in Figure 0.10.
2. If the data file was saved in a different location, continue to click on ▾ until the name of the appropriate location appears, and then click on this location name.
3. Next double-click on **PCDataSets,** and then double-click on **SPSS.**

4. Click on the name of the file you wish to open in SPSS for Windows. The desired file name will appear in the "File name" box. In this example, as shown in Figure 0.11, the name of the desired Microsoft Excel file is **ta01-06.xls.** SPSS will then convert the Excel file to an SPSS file. Follow the instructions given on-screen if you need to open a file that is not in SPSS (.por or .sav) format.

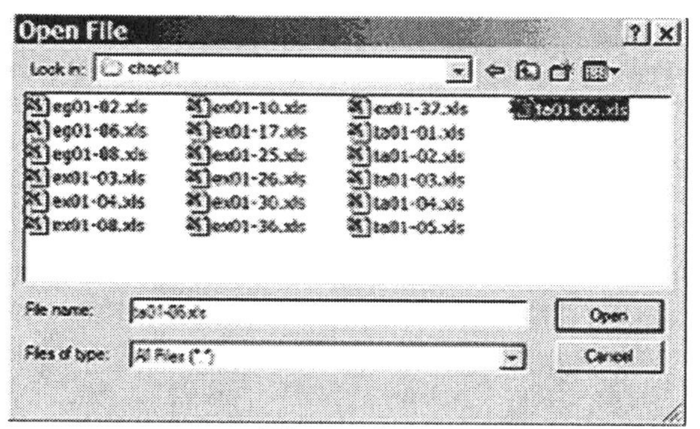

Figure 0.11

5. Click **Open.** The "Opening Excel Data Source" window appears, as shown in Figure 0.12. If the data set contained the variable names in the first row, click on the "Read variable names" box so that a check mark appears in the box. For this example, the variable names appear in the first row of the data set (see Figure 0.13); therefore, this box should be checked.

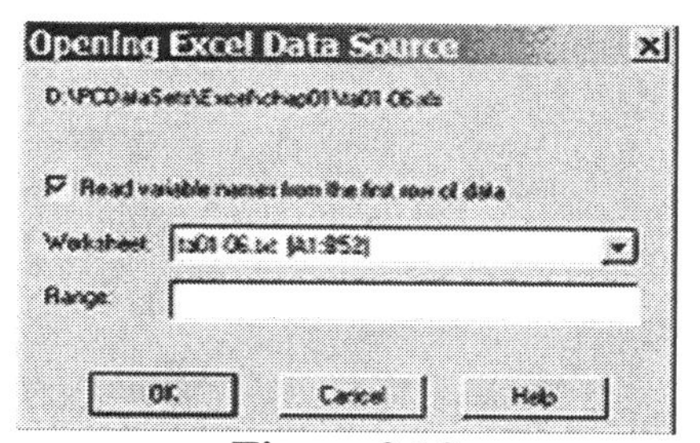

Figure 0.12

6. Click **OK.**
7. After the data are read into SPSS for Windows, the Output window of SPSS for Windows becomes the active window, and it gives a log containing the name of each variable read into SPSS for Windows, as well as the type and format of each variable (see Figure 0.13). Note that the first two variables in the data set are formatted as string 13, which means that the values are nonnumeric text with a width of 13 characters. The remaining variable is formatted as numeric 11.0, which means it is 11 digits wide with no digits for decimal places.

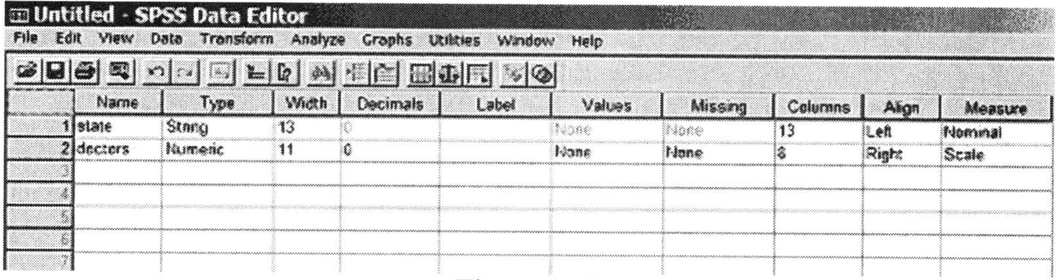

	Name	Type	Width	Decimals	Label	Values	Missing	Columns	Align	Measure
1	state	String	13	0		None	None	13	Left	Nominal
2	doctors	Numeric	11	0		None	None	8	Right	Scale
3										
4										
5										
6										
7										

Figure 0.13

8. To go to the actual data contained in the Data Editor, click **Window,** then click **1 Untitled - SPSS for Windows Data Editor.** Figure 0.14 shows both variables for the first 10 cases.

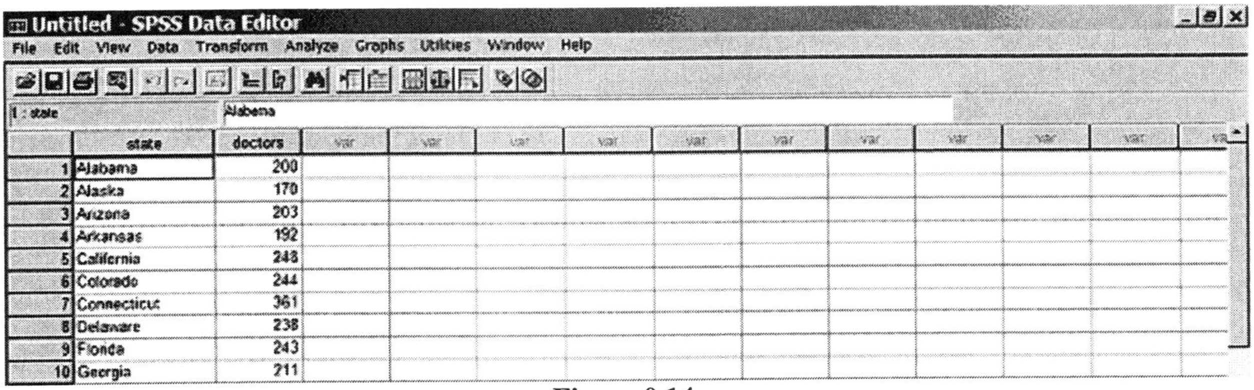

Untitled - SPSS Data Editor

File Edit View Data Transform Analyze Graphs Utilities Window Help

1 : state Alabama

	state	doctors	var	var	var	var	var	var	var	var	var	var	var
1	Alabama	200											
2	Alaska	170											
3	Arizona	203											
4	Arkansas	192											
5	California	248											
6	Colorado	244											
7	Connecticut	361											
8	Delaware	238											
9	Florida	243											
10	Georgia	211											

Figure 0.14

Defining a Variable

You might be interested in changing the default SPSS for Windows variable names to more appropriate variable names. **Variable names** in SPSS for Windows can be at most 8 characters long, containing no embedded blanks. However, you can add a **variable label,** a descriptive explanation of the variable name. Variable labels can be at most 256 characters long and can contain embedded blanks. The use of the variable label can make the SPSS for Windows output more informative.

The example used in this section is a small sample based on real data for individuals who completed the Social Problem Solving Inventory – Short Form (SPSI-SF). See Figure 0.15 for the data. The subject numbers have been changed to ensure anonymity. The data have already been typed into SPSS for Windows.

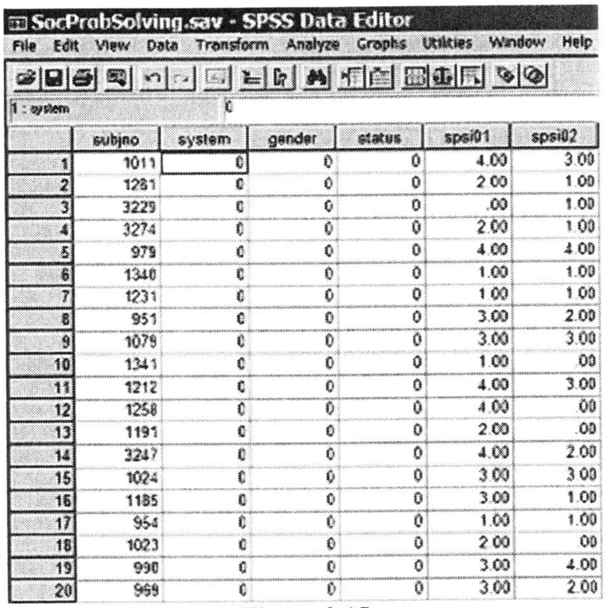

SocProbSolving.sav - SPSS Data Editor

File Edit View Data Transform Analyze Graphs Utilities Window Help

1 : system 0

	subjno	system	gender	status	spsi01	spsi02
1	1011	0	0	0	4.00	3.00
2	1281	0	0	0	2.00	1.00
3	3229	0	0	0	.00	1.00
4	3274	0	0	0	2.00	1.00
5	979	0	0	0	4.00	4.00
6	1340	0	0	0	1.00	1.00
7	1231	0	0	0	1.00	1.00
8	951	0	0	0	3.00	2.00
9	1079	0	0	0	3.00	3.00
10	1341	0	0	0	1.00	.00
11	1212	0	0	0	4.00	3.00
12	1258	0	0	0	4.00	.00
13	1191	0	0	0	2.00	.00
14	3247	0	0	0	4.00	2.00
15	1024	0	0	0	3.00	3.00
16	1185	0	0	0	3.00	1.00
17	954	0	0	0	1.00	1.00
18	1023	0	0	0	2.00	.00
19	990	0	0	0	3.00	4.00
20	969	0	0	0	3.00	2.00

Figure 0.15

To define a variable in SPSS for Windows (such as declaring the variable name, declaring the variable type, adding variable labels and value labels, and declaring the measurement type), follow these steps (the steps use the variable ***system*** as an example):

1. Click on the "Variable View" tab at the bottom left of the data set. Figure 0.16 shows the process of adding the labels "Male" and "Female" to ***gender.*** The value "0 = Male" has already been added. To complete the value labels for ***gender,*** enter the value "1" in the "Value" box and the label Female in the "Value Label" box and then click **Add,** then **OK.**

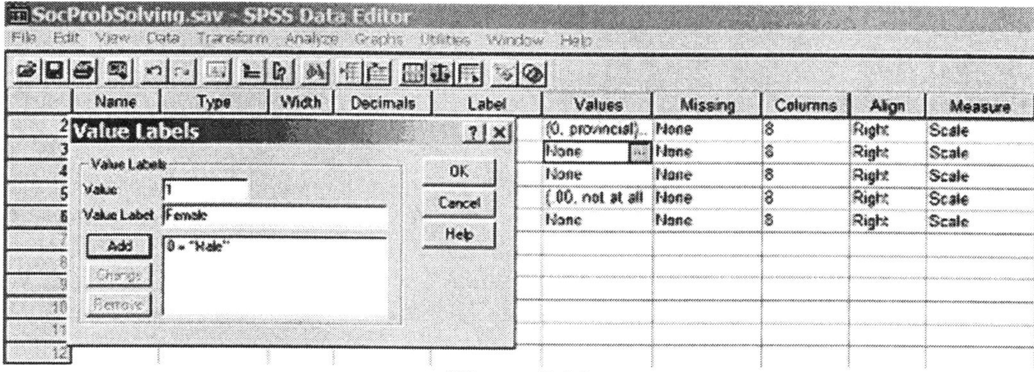

Figure 0.16

2. The value labels also have been added for ***spsi01*** and they are shown in Figure 0.17.

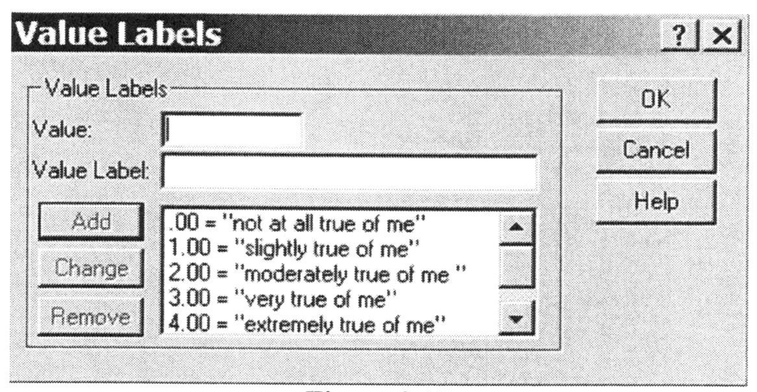

Figure 0.17

3. Select the appropriate measurement from the options available in the "Measurement" box: scale, ordinal, or nominal. The default option for numeric variables is scale.
4. If you are interested in adding additional information about the characteristics of your variables, click in the appropriate box and follow the instructions on the screen.

Recoding a Variable

Some of the analyses to be performed in SPSS for Windows will require that a categorical variable be entered into SPSS for Windows as a numeric variable (e.g., ***gender*** from the previous illustration). If the variable has already been entered as a string into the SPSS for Windows Data Editor, you can easily create a new variable that contains the same information as the string variable but is numeric simply by recoding the variable. In the example that follows, I have changed the numeric values for ***system*** from "0" and "1" to "1" and "2." When you recode variables like this, use numbers or string variables that make sense for your data.

To recode a variable into a different variable (the recommended option), follow these steps:

1. Click **Transform,** click **Recode,** and then click **Into Different Variables.** The "Recode into Different Variables" window in Figure 0.18 appears. Enter a new variable name (no more than 8 characters) in the "Name" box. In this example, I have used ***sys rec*** as my new variable name (meaning ***system recoded***).
2. Click **Old and New Values.** The "Recode into Different Variables: Old and New Values" window, also shown in Figure 0.18, appears.
3. In the "Value" box within the "Old Value" box, type **0** (the value has to be entered into this box *exactly* as it appears in the SPSS for Windows Data Editor). In the "Value" box in the "New Value" box, type **1** (or whatever you wish to change it to). Click **Add.** "0→1" appears in the "Old→ New" box. In the "Value" box within the "Old Value" box, type **1** (the value has to be entered into this box *exactly* as it appears in the SPSS for Windows Data Editor). In the "Value" box in the "New Value" box, type **2.** Click **Add.** "1→2" will now appear in the "Old→ New" box below the "0→1." If you have more than 2 levels of your variable, continue with this process until all have been assigned a unique numerical (or string if that is more appropriate) value. Remember that each time the value is entered into the "Old Value" box it must appear *exactly* as it appears in the SPSS for Windows Data Editor. When all values have been recoded, click **Continue.**
4. Click **OK.** The new variable ***sys rec*** appears in the SPSS for Windows Data Editor.

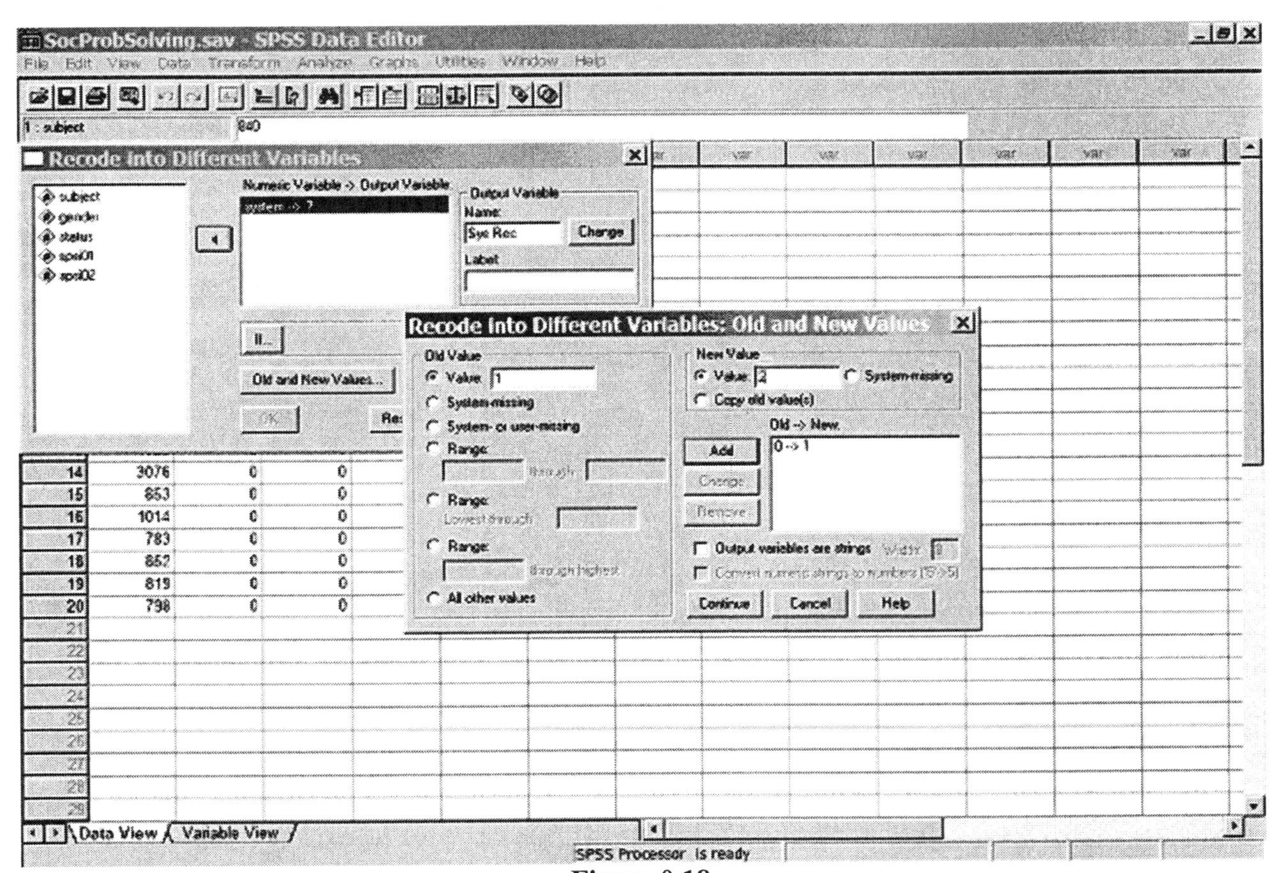

Figure 0.18

Deleting a Case from an SPSS for Windows Data File

Occasionally, extreme values (outliers) will be contained in data sets. It may be desirable to generate analyses both with and without the outlier(s) to assess the effect it (they) may have. Visual displays such as Stemplots, Histograms, and Bar Graphs (See Chapter 1) help you to understand your data and determine if a case should be deleted for your analysis. In what follows, I have assumed that you already know the number of the case you want deleted. To remove case 19 from the SPSS for Windows data file, follow these steps:

1. Use the scroll bar along the left side of the SPSS for Windows Data Editor so the desired case number appears among the numbers in the left-most, gray-colored column. Click on the desired case number. The entire row associated with that case number is highlighted (see Figure 0.19).
2. Click **Edit** and then click **Clear.** The row for that case number will be deleted. NOTE: Unless you want this case to be permanently deleted, **DO NOT SAVE THE DATA SET** upon exiting SPSS for Windows or the information for this case will be lost.

SocProbSolving.sav - SPSS Data Editor

File Edit View Data Transform Analyze Graphs Utilities Window Help

19 : subject 819

	subject	system	gender	status	spsi01	spsi02	var	var	var	var	var	var	var
1	840	0	0	0	4.00	3.00							
2	1110	0	0	0	2.00	1.00							
3	3058	0	0	0	.00	1.00							
4	3103	0	0	0	2.00	1.00							
5	808	0	0	0	4.00	4.00							
6	1169	0	0	0	1.00	1.00							
7	1060	0	0	0	1.00	1.00							
8	780	0	0	0	3.00	2.00							
9	908	0	0	0	3.00	3.00							
10	1170	0	0	0	1.00	.00							
11	1041	0	0	0	4.00	3.00							
12	1087	0	0	0	4.00	.00							
13	1020	0	0	0	2.00	.00							
14	3076	0	0	0	4.00	2.00							
15	853	0	0	0	3.00	3.00							
16	1014	0	0	0	3.00	1.00							
17	783	0	0	0	1.00	1.00							
18	852	0	0	0	2.00	.00							
19	819	0	0	0	3.00	4.00							
20	796	0	0	0	3.00	2.00							
21													
22													
23													

Figure 0.19

3. *NOTE:* With large data sets, you can also go to the desired case number by clicking **Data** and then clicking **Go to case.** The "Go to Case" window appears (see Figure 0.20). Type in the desired case number, click **OK,** and then click **Close.** The first cell of that case number will be highlighted. You will then need to click on the case number in the left-most, gray-colored column.

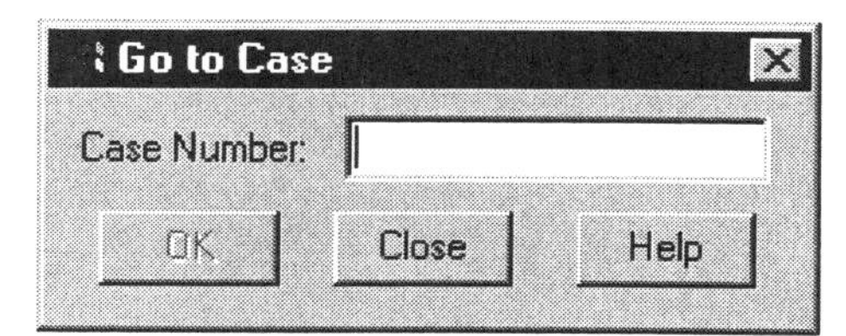

Figure 0.20

These steps can be repeated if you want to delete other cases. CAUTION: Once an observation is deleted, SPSS for Windows renumbers all subsequent cases. Therefore, the case number for a subsequent case will change once an earlier case number has been deleted.

Opening Data Sets Not Created by SPSS or Windows Excel

Although many types of data files can be read directly into SPSS for Windows, I recommend initially reading the data file into Microsoft Excel and then reading the Microsoft Excel file into SPSS for Windows. The following example illustrates the conversion process using the data associated with Table 1.3 in BPS, which is contained on the CD-ROM provided with BPS. In this example, I will import data in a .dat format. This type of data may have been produced with a word processor such as Word for Windows.

To convert a text data file into a Microsoft Excel file, follow these steps:

1. Open Microsoft Excel.
2. Click **File** and then click **Open.** The "Open" window appears.
3. Click ▾ in the "Look in" box until you find the name of the drive that contains the text data file, and then click on this name. For this example, the drive that contains the ASCII text data file of interest is drive **D:.** Click ▾ in "Files of type" box, and then click **All Files [*.*].**
4. Click on **PCDataSets, Text, chap01, ta01-03.dat.** The "Text Import Wizard" window opens, as shown in Figure 0.21. The "Text Import Wizard" determines that the data are of fixed width (variable values are lined up in columns). As can be seen in the "Start import at row" box, the user can set the "Text Import Wizard" to begin importing the data at row 2, which is correct (because variable names appear as column headings in the .dat file).

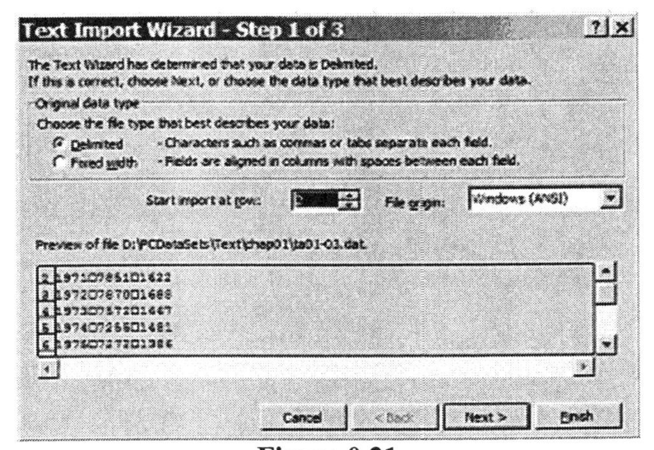

Figure 0.21

5. Click **Next>.** The second window of the "Text Import Wizard" appears, as shown in Figure 0.22. Each column of numbers in the "Data preview" box represents the different variables in the data set. The "Text Import Wizard" suggests where the column breaks should occur between the variables. Additional column breaks can be inserted by clicking at the desired location within the "Data preview" box. An existing column break can be removed by double-clicking on that column break. The breaks suggested by the "Text Import Wizard" are appropriate for these data.

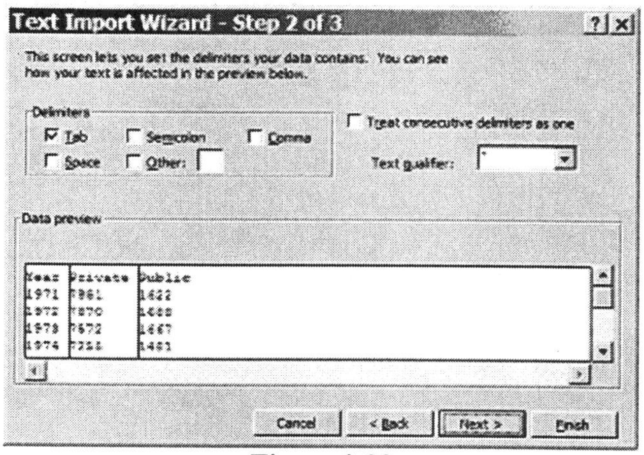

Figure 0.22

6. Click **Next>.** The third window of the "Text Import Wizard" appears, as shown in Figure 0.23. Here you can exclude variables, if desired, by clicking on the column you do not wish to import and selecting "Do not import column (Skip)" in the "Column data format" box.
You can also specify the format for each variable. By default, all variables are assigned a General format; this provides the most flexibility in SPSS for Windows.

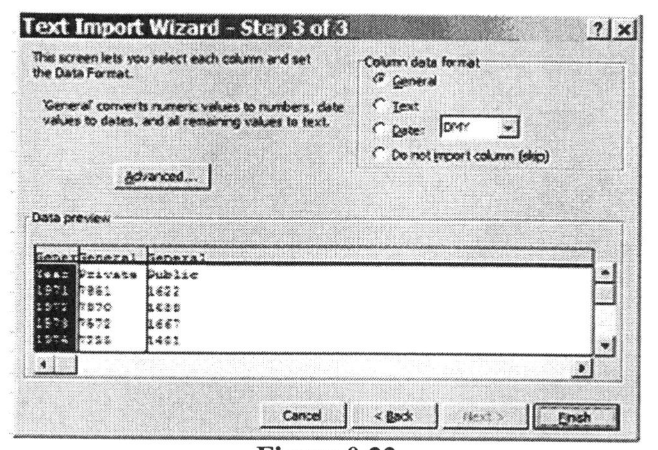

Figure 0.23

7. Click **Finish.** The data will appear in the Microsoft Excel spreadsheet, where the columns represent variables and the rows represent cases. Figure 0.24 shows the first 10 rows of the data set

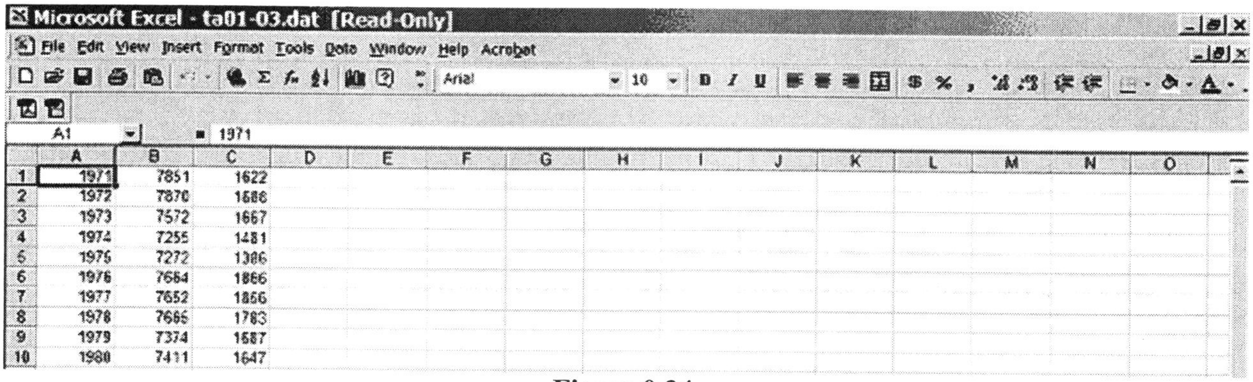

Figure 0.24

Printing in SPSS for Windows

Printing Data

To print out the data in the SPSS for Windows Data Editor, follow these steps:

1. Make the SPSS for Windows Data Editor the active window (click somewhere on the SPSS for Windows Data Editor).
2. Click **File,** then click **Print.** A "Print" window similar to the one shown in Figure 0.25 will appear.

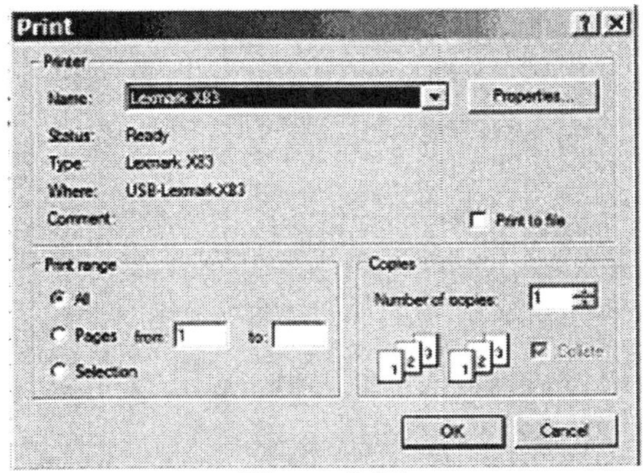

Figure 0.25

3. SPSS for Windows assumes that you want to print the entire data set. If you wish to print only a subset of the data (e.g., only one variable), you must first click on the appropriate column heading(s). If this is done, "Selection" in Figure 0.25 rather than "All" will be highlighted.
4. Click **OK.**

Printing Output and Charts in SPSS for Windows

To print out SPSS for Windows output and SPSS for Windows charts, follow these steps:

1. Make the "Output – SPSS Viewer" window the active window (see Figure 26).

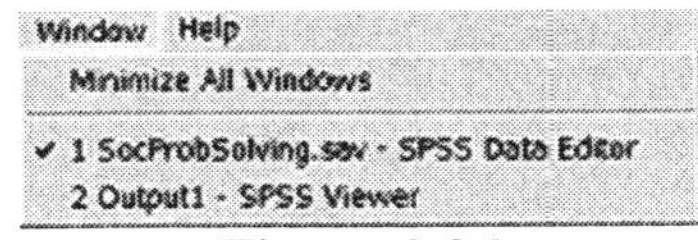

Figure 0.26

2. If you want to print all of the output, including graphs, click the Output icon (Output), which appears at the top of the left-hand window. All output is highlighted, as shown in Figure 0.27.

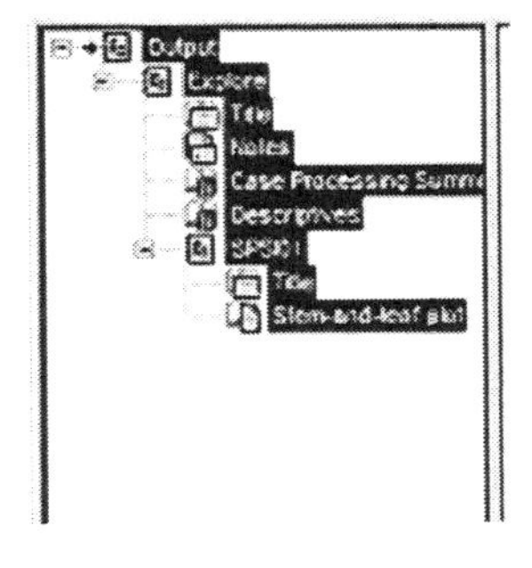

Figure 0.27

3. If you want to print only a specific portion of the output, such as only the Stem-and leaf, click on that icon (see Figure 0.28).

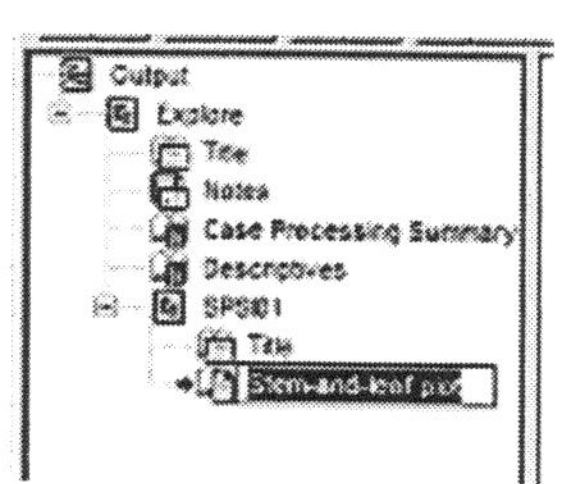

Figure 0.28

4. Click **File,** click **Print,** and then click **OK.**
5. After the output has been printed, it can be deleted. This is accomplished by first highlighting the portion of the output you wish to delete. For example, you can highlight all of the output, as shown in Figure 0.27, or highlight a specific icon, as shown in Figure 0.28. Then after the desired output has been selected, you can delete the output by pressing the **Delete** key.

Copying from SPSS for Windows into Microsoft Word 97

Copying a Chart or Table

To copy a chart or table from SPSS for Windows into Microsoft Word 97, follow these steps (note: It is assumed that Microsoft Word 97 has already been opened):

1. In the "Output – SPSS for Windows Viewer" window, select the chart or table to be copied by clicking on the icon that appears in the left side of the window (for example, the **Stem-and-leaf** icon shown in Figure 0.29 or Frequencies shown in Figure 0.30).

Figure 0.29

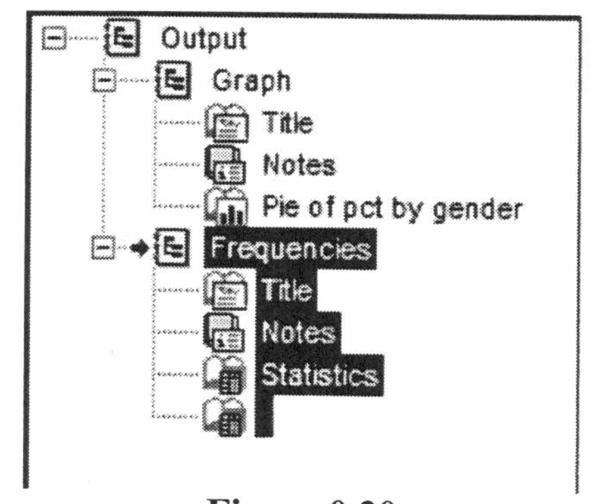

Figure 0.30

2. Click **Edit** and then click **Copy objects.**
3. In Microsoft Word 97, position the cursor at the desired place in the document, click **Edit** and then click **Paste.**
4. From the Microsoft Word for Windows Main Menu, click **Format**, and then click **Picture.**

5. A "Format Picture" window appears. Click on the **Position** tab.
6. Click **Float over text**. The ✔ in front of "Float over text" disappears. For other versions of Microsoft Word, explore the advanced options.
7. Click **OK.**
8. If you are interested in resizing the chart, click on the chart and place the cursor on one of the little black squares ("handles") that appear along the box that now outlines the chart. Using the right mouse button, drag the cursor in or out depending on how you want to resize the chart. Using one of the four corner squares will resize the chart but maintain the scale of width to height. The two center squares along the sides will change the width of the chart but maintain the current height. The two center squares along the top and bottom will change the height of the chart but maintain the current width. Care should be taken in resizing a chart because it is easy to distort the chart's overall image and the information the chart is meant to relay.
9. If you have chosen a table, it will appear in Microsoft Word for Windows like Table 0.1.
10. To begin typing in Microsoft Word 97, click on any area within the document that is outside the chart or table.

Frequencies

Statistics

N	Valid	36
	Missing	0

		Frequency	Percent	Valid Percent	Cumulative Percent
Valid	F	24	66.7	66.7	66.7
	M	12	33.3	33.3	100.0
	Total	36	100.0	100.0	

Table 0.1

Using SPSS for Windows Help

SPSS for Windows has extensive and useful on-line help. Suppose you want to know the steps needed to obtain a boxplot using SPSS for Windows and no manual is available. This information can be obtained using the help function of SPSS for Windows by following these steps:

1. Click **Help,** and then click **Topics.** The "Help Topics" window appears, as shown in Figure 0.31.

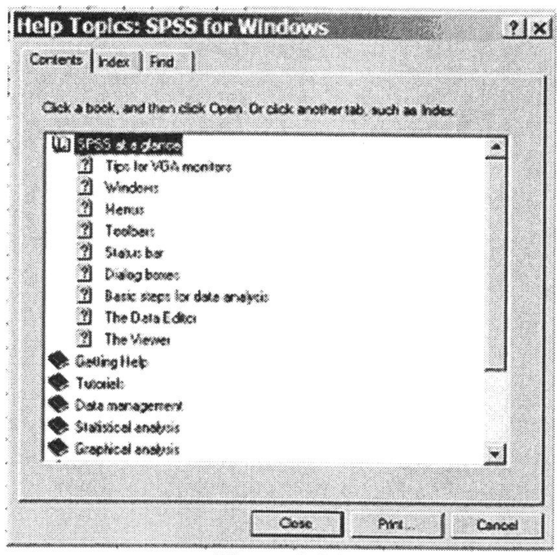

Figure 0.31

2. Click the **Index** tab and type **Box** in the "Type the first few letters of the word you're looking for" box. Figure 0.32 shows the list of potential terms from which to choose.

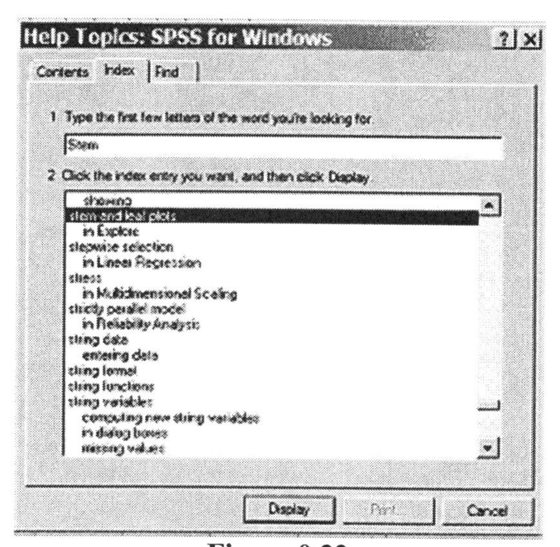

Figure 0.32

3. Double-click in **Explore** under "stem and leaf plots." The directions titled "To Obtain Simple and Clustered Boxplots" appear in the "How To" window (see Figure 0.33).

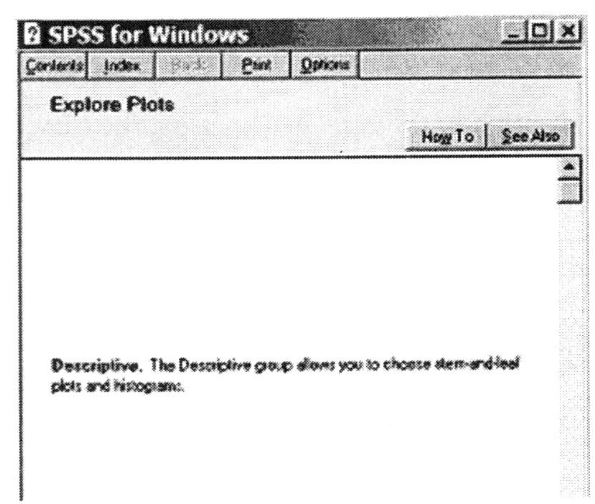

Figure 0.33

4. To print the directions, click **Print**.
5. To exit SPSS for Windows Help, click in the upper right corner of the "SPSS for Windows" window (see Figure 0.33).

Postscripts

I recommend that you develop a record-keeping system so that you will be able to find the files that you create in the future. I have included a possible chart below.

FILE NAME AND LOCATION	CHAPTER AND EXERCISE NUMBER REFERENCE	BRIEF DESCRIPTION OF THE DATA

Just for Fun!

Now that you have done all this hard work, for an enjoyable hour or two, stop to read *The Number Devil: A Mathematical Adventure*[1].

[1] Enzensberger, H. M. *The Number Devil: A Mathematical Adventure.* New York: Metropolitan Books; Henry Holt and Company, 1997.

Chapter 1. Picturing Distributions with Graphs

Topics covered in this chapter:

- **Pie Charts**
- **Bar Graphs or Bar Charts**
- **Histograms**
- **Stemplots**
- **Time plots**

Statistical tools and ideas help us examine data. This examination is called exploratory data analysis. This section introduces the notion of using graphical displays to perform exploratory data analysis. The graphical display used to summarize a single variable depends on the type of variable being studied (i.e., whether the variable is categorical or quantitative). For categorical variables, **pie charts** and **bar graphs** will be examined. For quantitative variables, **histograms, stemplots,** and **time plots** will be examined.

The data shown in Example 1.1 will be used to illustrate pie charts and bar graphs. These data are categorical. To make your own example, open a blank SPSS for Windows Data Editor screen (refer to Chapter 0 for a review of how to do this) and enter the data in a string variable labeled ***vote.***

Example 1.1
Twenty randomly selected faculty members were asked whether they would vote in favor of the new General Education requirements proposed to be implemented within the next two years. The data are given in Table 1.1.

Table 1.1 Votes on General Education Requirements Proposal

Yes	Yes	No	Undecided	No
Yes	No	Yes	Undecided	Yes
Yes	No	Yes	Undecided	No
No	No	Yes	Undecided	Undecided

Pie Charts

To create a pie chart for a categorical variable (such as ***vote*** in Example 1.1 above), follow these steps:

1. Click **Graphs** and then click **Pie.** The "Pie Charts" window in Figure 1.1 appears.

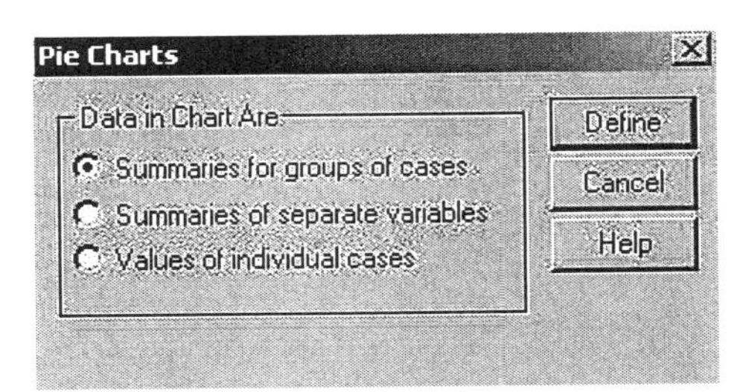

Figure 1.1

2. Click **Define.** The "Define Pie: Summaries for Groups of Cases" window in Figure 1.2 appears.

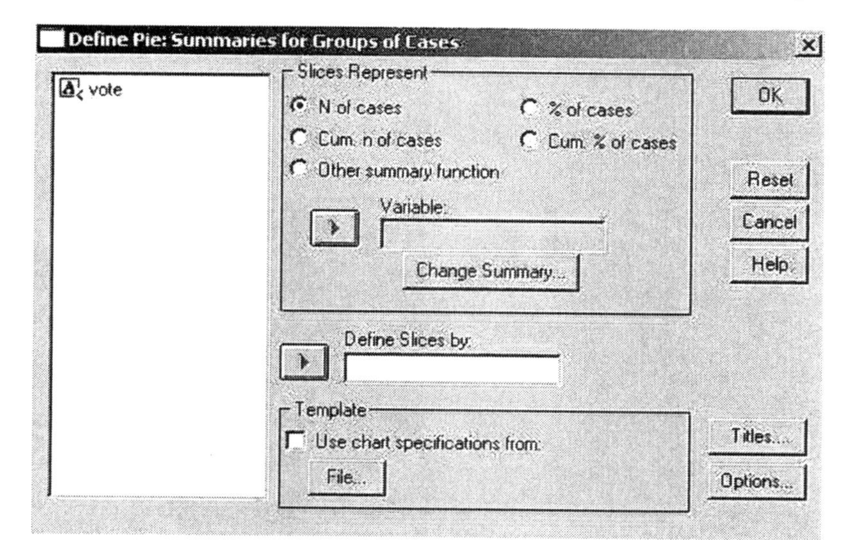

Figure 1.2

3. Click ***vote,*** then click ▸ to move ***vote*** into the "Define Slices by" box.
4. By default, the slices represent the number of cases. The same pie chart will appear if the **% of cases** within the "Slices Represent" box is selected. For this reason, the default option is recommended.
5. If you are interested in including a title or a footnote on the chart, click **Titles.** The "Titles" window shown in Figure 1.3 appears. In the properly labeled box ("Title" or "Footnote"), type in the desired information. Click **Continue.**
6. Click **OK.**

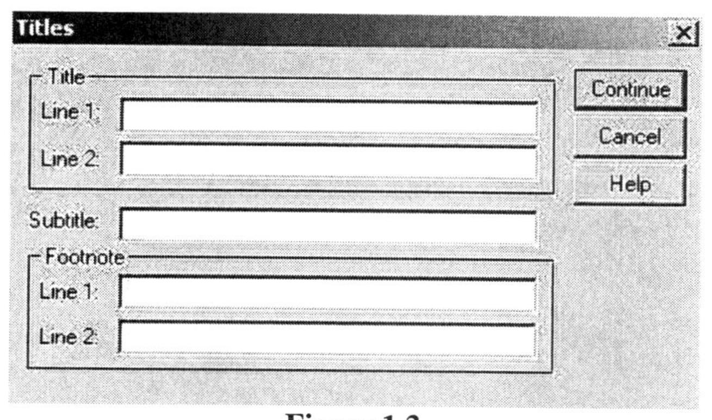

Figure 1.3

Figure 1.4 on the following page is the resulting SPSS for Windows output (except for the difference in the color scheme), after having selected **N of cases** rather than **% of cases** in the "Slices Represent" box.

Editing Pie Charts

By default, only the category labels of the categorical variable appear on the pie chart. To add the counts and percentages to the pie slices, follow these steps:

1. Double-click on the pie chart in the "Output – SPSS for Windows Viewer" window. The pie chart

now appears in the "Chart – SPSS for Windows Chart Editor" window, which has new menu and tool bars. The chart in the "Output – SPSS for Windows Viewer" window will be shaded with black diagonal lines whenever the chart is in the "Chart – SPSS for Windows Chart Editor" window.

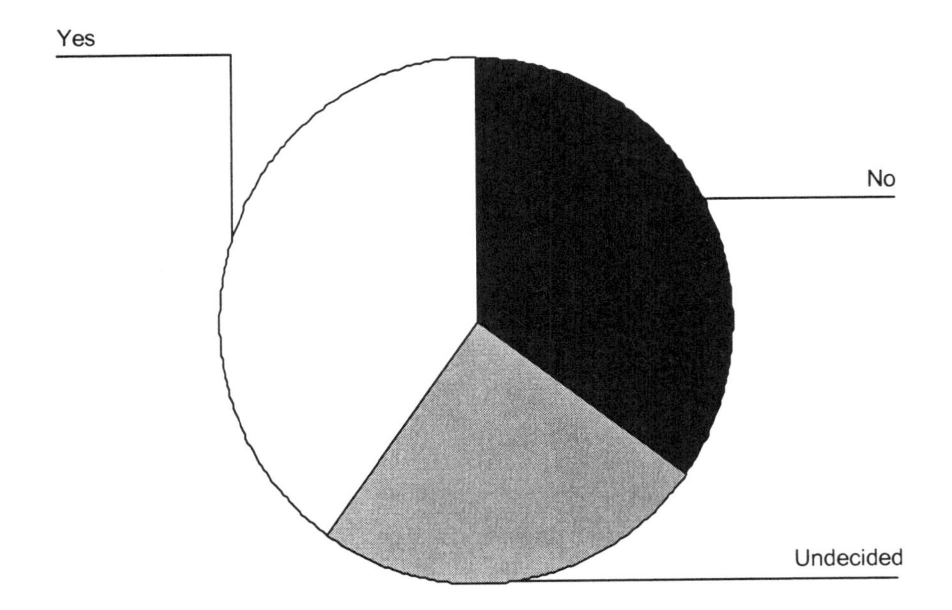

Figure 1.4

2. Click **Chart** from the main menu bar within the "Chart – SPSS for Windows Chart Editor" window and then click **Options.** The "Pie Options" window in Figure 1.5 appears.

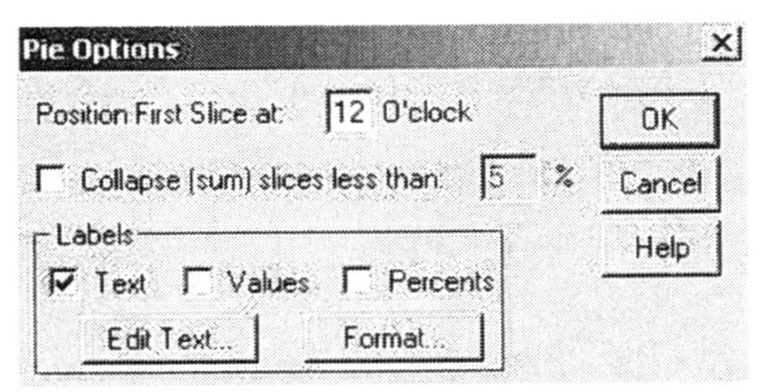

Figure 1.5

3. In the "Labels" box, click **Values** and then click **Percents**. The counts and percents appear on the outside of the pie chart beside the category text (for reference, see Figure 1.4). If this is the desired format for the output, skip to step 7.
4. If you would like the numbers to appear inside the slices, click **Format.** The "Pie Options: Label Format" window in Figure 1.6 appears.

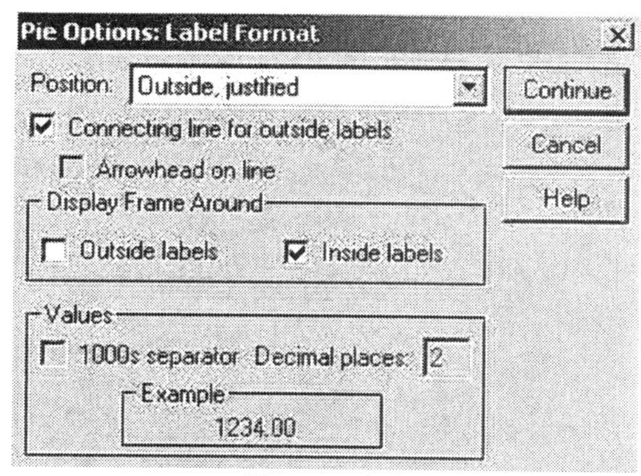

Figure 1.6

5. Click ▼ located next to the "Position" box until **Numbers inside, text outside** is highlighted.
6. Click **Continue.**
7. Click **OK.**
8. If you are finished editing the chart, click **File** from the main menu bar within the "Chart – SPSS for Windows Chart Editor" window and then click **Close** to return to the "Output – SPSS for Windows Viewer" window.

Figure 1.7 is the resulting SPSS for Windows output after adding the counts and percents inside the pie slices and changing the colors of the slices. To change the color or the pattern-fill of a pie slice, follow the directions about changing the color or the fill in the Editing Bar Graphs section later in this chapter (see Figures 1.12 and 1.13).

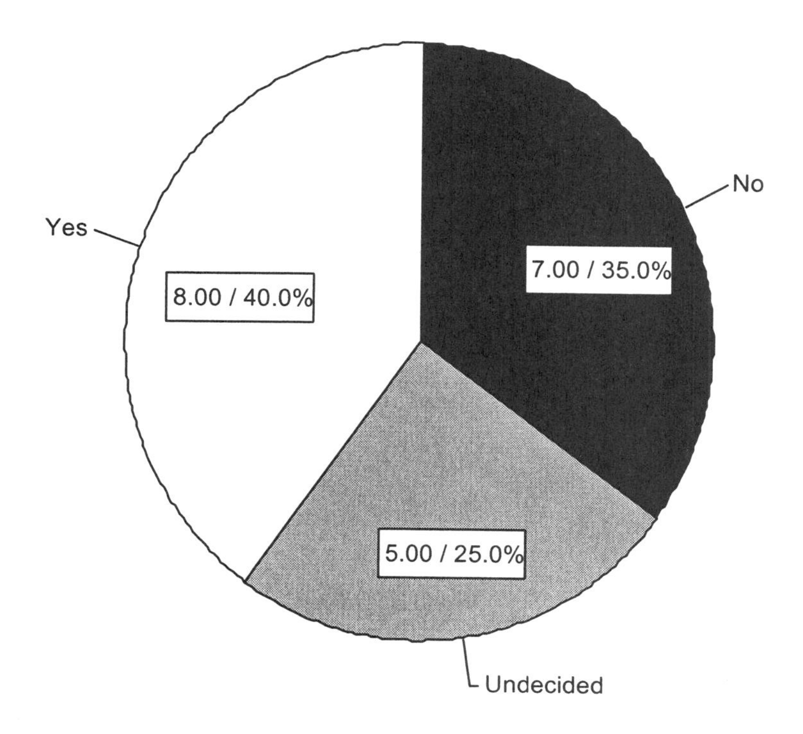

Figure 1.7

Bar Graphs or Bar Charts

To create a bar chart for a categorical variable (such as ***vote*** in Example 1.1 above), follow these steps:

1. Click **Graphs** and then click **Bar.** The "Bar Charts" window in Figure 1.8 appears.

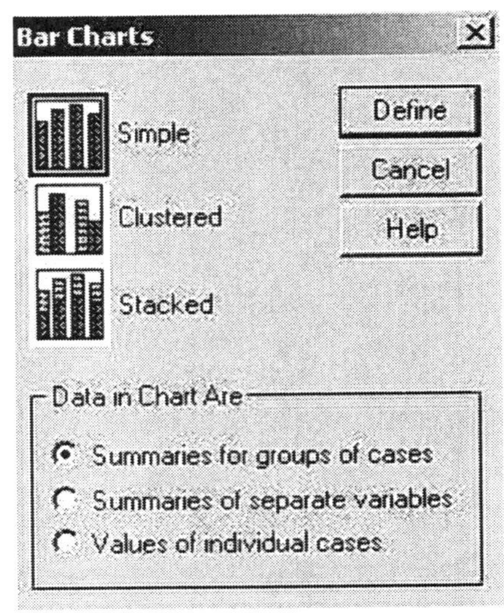

Figure 1.8

2. Click **Define.** The "Define Simple Bar: Summaries for Groups of Cases" window in Figure 1.9 appears.

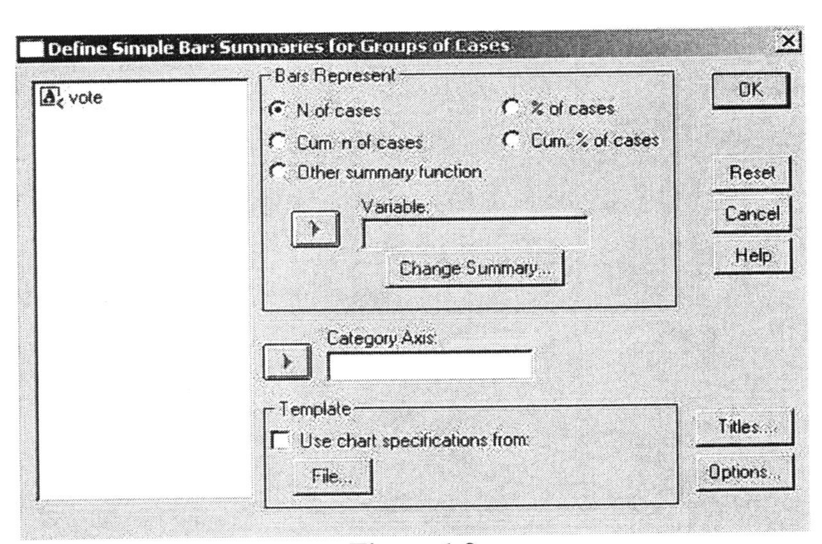

Figure 1.9

3. Click ***vote,*** then click ‣ to move ***vote*** into the "Category Axis" box.
4. By default, the bars represent the counts (N of cases). If you are interested in the *y* axis being labeled as "Percent" rather than "Count," click **% of cases** in the "Bars Represent" box.

5. If you are interested in including a title or a footnote on the chart, click **Titles.** The "Titles" window in Figure 1.3 appears. In the properly labeled box ("Title" or "Footnote"), type in the desired information. Click **Continue.**
6. Click **OK.**

Figure 1.10 is the resulting SPSS for Windows output.

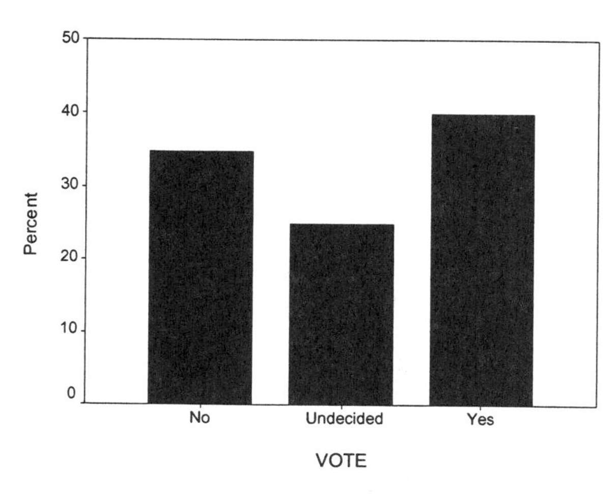

Figure 1.10

Editing Bar Charts

To have the numbers appear within the bars, follow these steps:

1. Double-click on the bar chart in the "Output – SPSS for Windows Viewer" window. The bar chart now appears in the "Chart – SPSS for Windows Chart Editor" window, which has new menu and tool bars. The chart in the "Output – SPSS for Windows Viewer" window will be shaded with black diagonal lines whenever the chart is in the "Chart – SPSS for Windows Chart Editor" window.
2. Click [icon]. The "Bar Label Styles" window in Figure 1.11 appears.

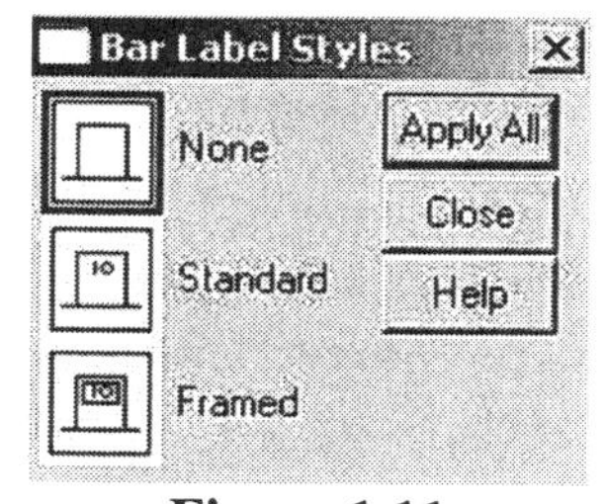

Figure 1.11

3. Click on the box in front of the **Framed** option. "None" is the default option.
4. Click **Apply All.**
5. Click **Close.**
6. If you are finished editing the chart, click **File** from the main menu bar within the "Chart – SPSS for Windows Chart Editor" window and then click **Close** to return to the "Output – SPSS for Windows

Viewer" window.

To change the color of an area within the chart (or to change the color of a pie slice), follow these steps:

1. The chart must be in the "Chart – SPSS for Windows Chart Editor" window. If the chart is not in the "Chart – SPSS for Windows Chart Editor" window, double-click on the chart in the "Output – SPSS for Windows Viewer" window.
2. Click on the area within the chart for which a change in color is desired (for instance, the bars in the bar chart). The area to be changed will be outlined with little black squares. For this example, the three bars in the chart are outlined in little black squares.
3. Click [icon]. The "Colors" window in Figure 1.12 appears.

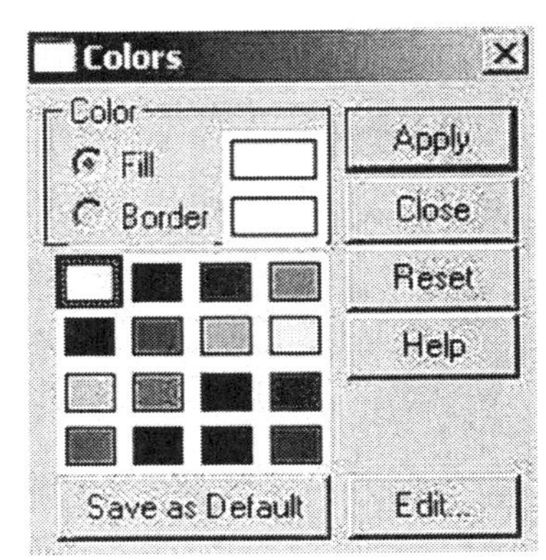

Figure 1.12

4. Make sure that "Fill" rather than "Border" is selected within the "Color" box. Click on the desired color for the fill for instance, light gray.
5. Click **Apply.**
6. Click **Close.**
7. If you are finished editing the chart, click **File** from the main menu bar within the "Chart – SPSS for Windows Chart Editor" window and then click **Close** to return to the "Output – SPSS for Windows Viewer" window.

To change the pattern of an area within the chart, follow these steps:

1. The chart must be in the "Chart – SPSS for Windows Chart Editor" window. If the chart is not in the "Chart – SPSS for Windows Chart Editor" window, double-click on the chart in the "Output – SPSS for Windows Viewer" window.
2. Click on the area within the chart for which a change in pattern is desired (for instance, the bars in the bar chart). The area to be changed will be outlined with little black squares. For this example, the three bars in the chart are outlined in little black squares.
3. Click [icon]. The "Fill Patterns" window in Figure 1.13 appears.

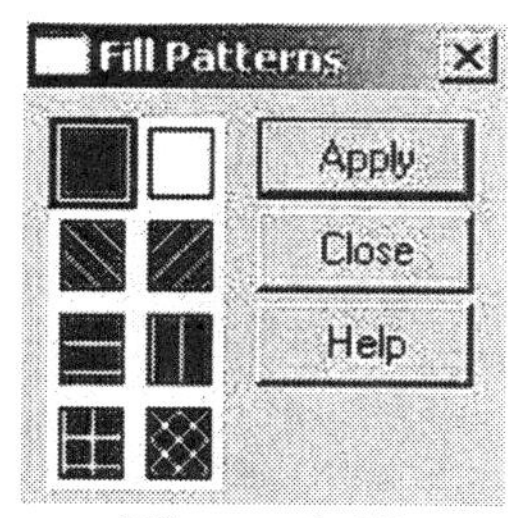

Figure 1.13

4. Click the desired pattern. A solid color pattern is the default option.
5. Click **Apply,** then click **Close.**
6. If you are finished editing the chart, click **File** from the main menu bar within the "Chart – SPSS for Windows Chart Editor" window and then click **Close** to return to the "Output – SPSS for Windows Viewer" window.

To make changes to the *x* axis (such as changing the axis title or the orientation of the labels), follow these steps:

1. The chart must be in the "Chart – SPSS for Windows Chart Editor" window. If the chart is not in the "Chart – SPSS for Windows Chart Editor" window, double-click on the chart in the "Output – SPSS for Windows Viewer" window.
2. Click **Chart** and then click **Axis.** The "Axis Selection" window in Figure 1.14 appears.

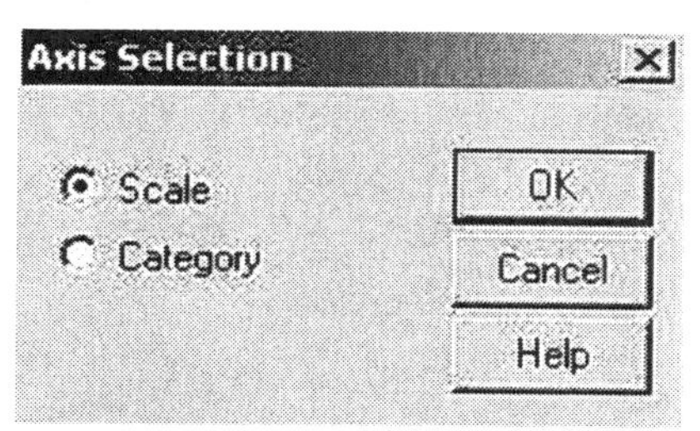

Figure 1.14

3. Click **Category** (for *x* axis) and then click **OK.** The "Category Axis" window in Figure 1.15 appears.

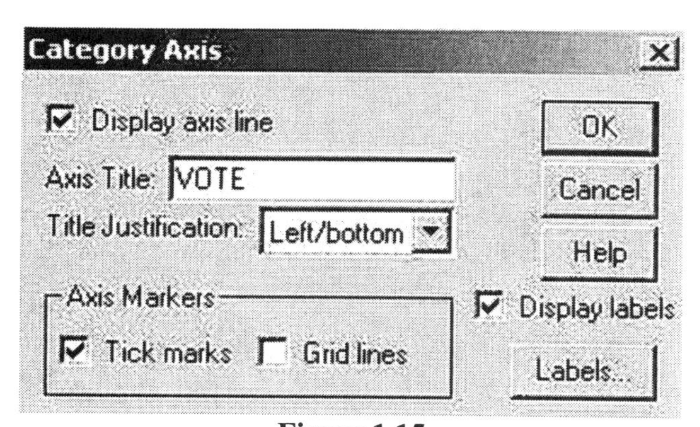

Figure 1.15

4. To change the title of the axis, replace VOTE in the "Axis Title" box with the desired axis title (for instance, **Voting Preference for General Education Requirements**).

5. To center the axis title, click ▾ in the "Title Justification" box and click **Center.**
6. To change the labels for the categorical variable appearing on the *x* axis, click **Labels.** The "Category Axis: Labels" window in Figure 1.16 appears.

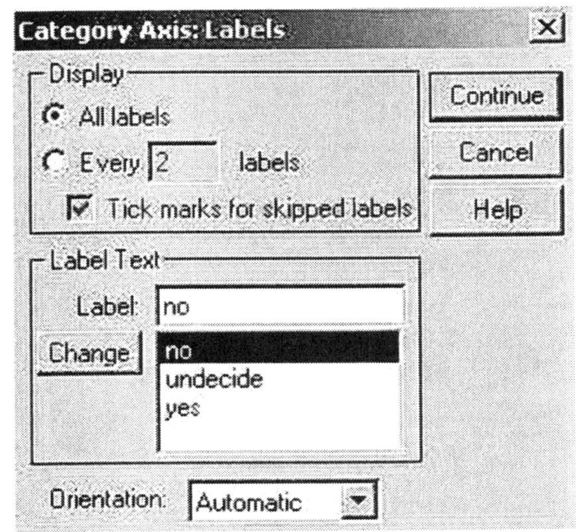

Figure 1.16

7. Type the desired label to replace "No" within the "Label" box and then click **Change.** Click **Undecided** within the "Label Text" box, type the desired label to replace "Undecided" within the "Label" box, and then click **Change.** Repeat this process for the rest of the labels. You are encouraged to assign the desired label values to the variable before creating charts for the variable (see Chapter 0 for labeling variables).
8. To change the orientation of the labels on the *x* axis, click ▾ in the "Orientation" box and then click the desired orientation style (for instance, **Diagonal**).
9. Click **Continue.**
10. Click **OK.**
11. If you are finished editing the chart, click **File** from the main menu bar within the "Chart – SPSS for Windows Chart Editor" window and then click **Close** to return to the "Output – SPSS for Windows Viewer" window.

To make changes to the *y* axis (such as changing the axis title, the range of values, or the spacing of the tick marks), follow these steps:

1. The chart must be in the "Chart – SPSS for Windows Chart Editor" window. If the chart is not in the "Chart – SPSS for Windows Chart Editor" window, and double-click on the chart in the "Output – SPSS for Windows Viewer" window.
2. Click **Chart** and then click **Axis.** The "Axis Selection" window as shown above in Figure 1.14 appears.
3. Click **Scale** (for *y* axis) and then click **OK.** The "Scale Axis" window in Figure 1.17 appears.

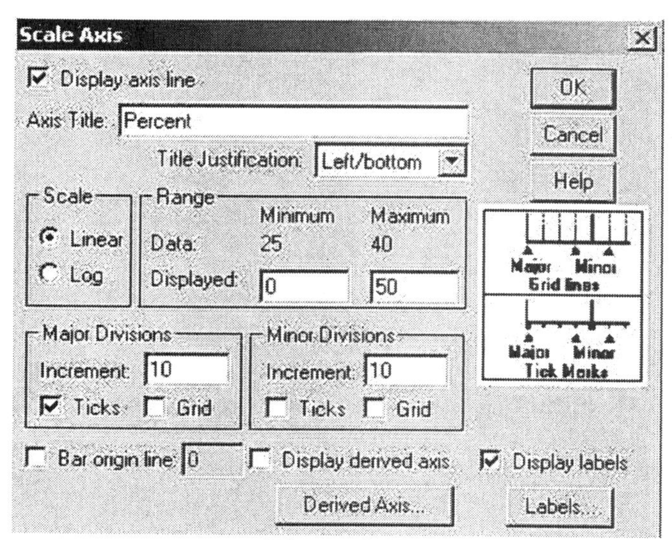

Figure 1.17

4. To change the title of the y axis, replace Percent in the "Axis Title" box with the desired label, such as **Percentage of Cases.**
5. To center the axis title, click ▼ in the "Title Justification" box and click **Center.**
6. To change the range of values displayed on the y axis, for instance to 0 to 45, type **0** in the "Minimum Displayed" box and type **45** in the "Maximum Displayed" box. Using 0 as the minimum value for most charts is advisable to avoid generating misleading graphs.
7. To change the spacing of the tick marks on the y axis, for instance to 5, type **5** in the "Major Divisions Increment" box. Note: The axis range must be an even multiple of the major increment.
8. Click **OK.**
9. If you are finished editing the chart, click **File** from the main menu bar within the "Chart – SPSS for Windows Chart Editor" window and then click **Close** to return to the "Output – SPSS for Windows Viewer" window.

Figure 1.18 is the resulting SPSS for Windows output after adding the numbers inside the bars, changing the color of the bars to light gray, and editing the x and y axes.

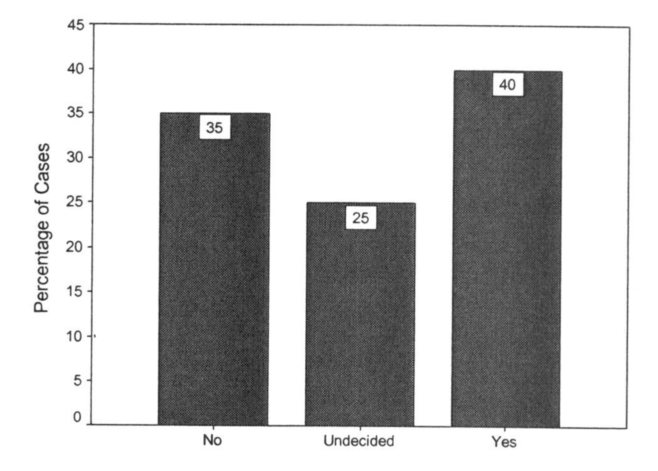

Figure 1.18

Histograms

A histogram breaks the range of values of a quantitative variable into intervals and displays only the count or the percentage of the observations that fall into each interval. You can choose any convenient number of intervals, but you should always choose intervals of equal width.

Example 1.3 and Table 1.1 in BPS:
Table 1.1 in BPS presents the percentage of residents in each of the 50 states who identified themselves in a 2000 census as "Spanish/Hispanic/Latino." These data can be retrieved from the file ***ta01-01.por*** in the SPSS data sets on your CD-ROM. When you open the file, the SPSS for Windows Data Editor contains two variables called ***state*** (declared type string 30) and ***percent*** (declared type numeric 8.1).

To create a frequency histogram of this distribution, follow these steps:

1. Click **Graphs** and then click **Histogram.** The "Histogram" window in Figure 1.19 appears.
2. Click ***percent,*** then click ▸ to move ***percent*** into the "Variable" box.
3. If you want to include a title or a footnote on the chart, click **Titles.** The "Titles" window shown earlier in Figure 1.3 appears. In the properly labeled box ("Title" or "Footnote"), type in the desired information.

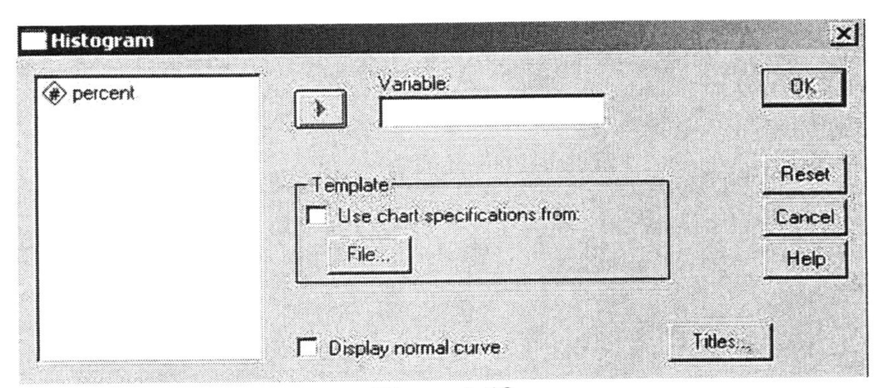

Figure 1.19

4. Click **Continue.**
5. Click **OK.**

Figure 1.20 is the default histogram created by SPSS for Windows. The class intervals for the default histogram are 4.5 to under 5.5, 5.5 to under 6.5, . . . , 18.5 to under 19.5.

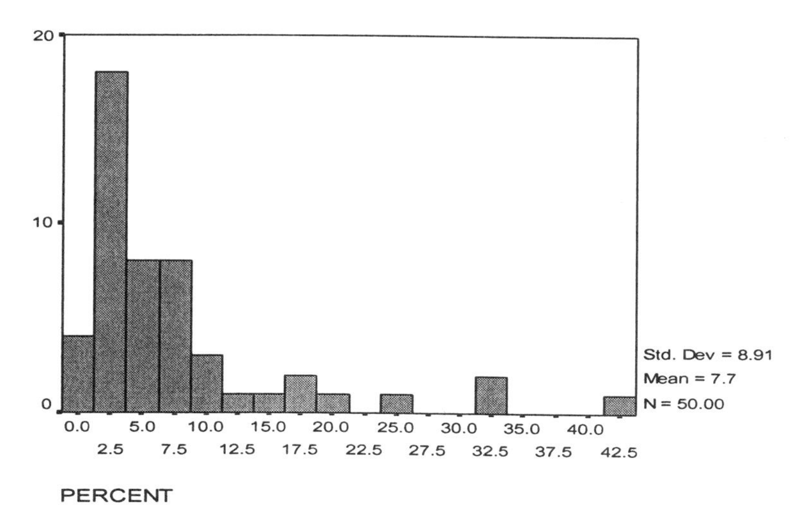

Figure 1.20

Editing Histograms

To obtain a picture of the distribution similar to that shown in Figure 1.3 of BPS, you can edit the histogram. A distinct difference between the SPSS for Windows default histogram and the histogram in Figure 1.3 of BPS is the chosen class intervals. The histogram in Figure 1.3 of BPS has the following class intervals: 0.0 to 4.9, 5.0 to 9.9, . . . , 40.0 to 44.9. To edit the histogram, double-click on the histogram in the "Output – SPSS for Windows Viewer" window. The histogram now appears in the "Chart – SPSS for Windows Chart Editor" window, which has its own menu and tool bars.

To make changes to the *x* axis (such as changing the title or the class width of the bars), follow these steps:

1. Click **Chart** and then click **Axis.** The "Axis Selection" window in Figure 1.21 appears.

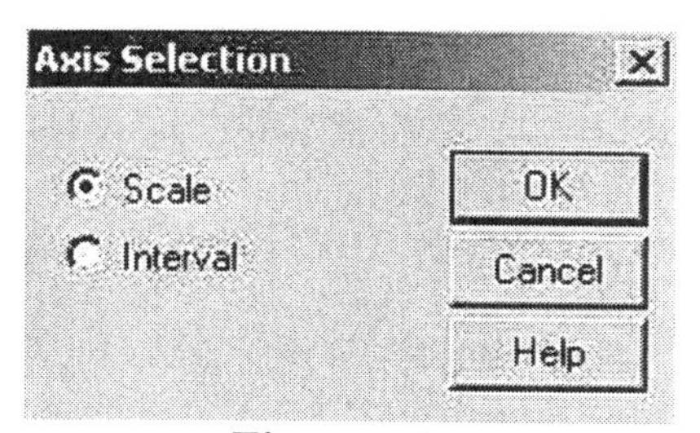

Figure 1.21

2. Click **Interval** (for the *x* axis) and then click **OK.** The "Interval Axis" window in Figure 1.22 appears.

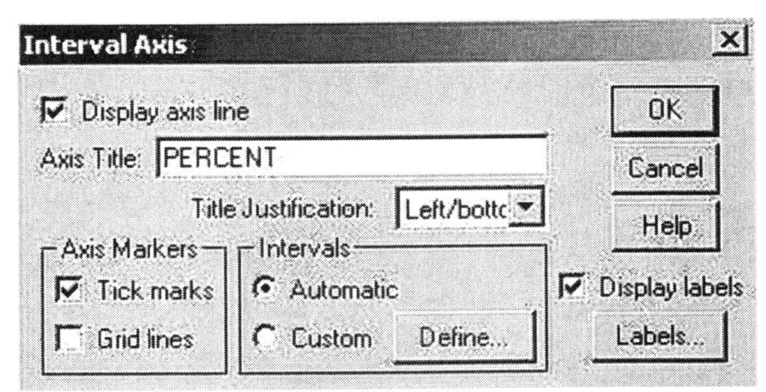

Figure 1.22

3. To change the title of the axis, replace PERCENT within the "Axis Title" box with the desired axis title (for instance, **Percent of Hispanic State Residents in 2000**).
4. To center the axis title, click ▼ in the "Title Justification" box and click **Center.**
5. To change the range of values on the *x* axis and/or the class width of the bars, click **Custom** and then click **Define** located within the "Intervals" box. The "Interval Axis: Define Custom In…" window in Figure 1.23 appears.

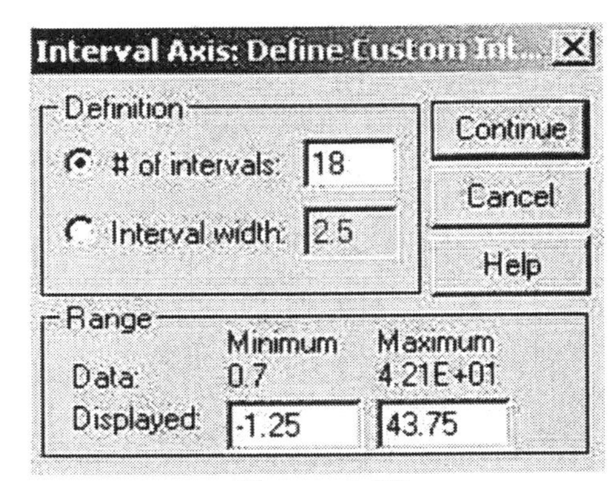

Figure 1.23

6. To change the class width, for instance to 5, click **Interval width** and type **5** in the "Interval width" box.
7. To change the range of the *x* axis, for instance 0.0 to 45.0, replace -1.25 with **0.0** in the "Minimum Displayed" box and replace 43.75 with **45.0** in the "Maximum Displayed" box. Note: Given the change in the range of the *x* axis and the interval width of 5, the new class intervals are 0.0 to 4.9, 5.0 to 9.9, . . . , and 40.0 to 44.9.
8. Click **Continue.**
9. To edit the number and the style of labels appearing on the *x* axis, click **Labels** in the "Interval Axis" window. The "Interval Axis: Labels" window in Figure 1.24 appears.

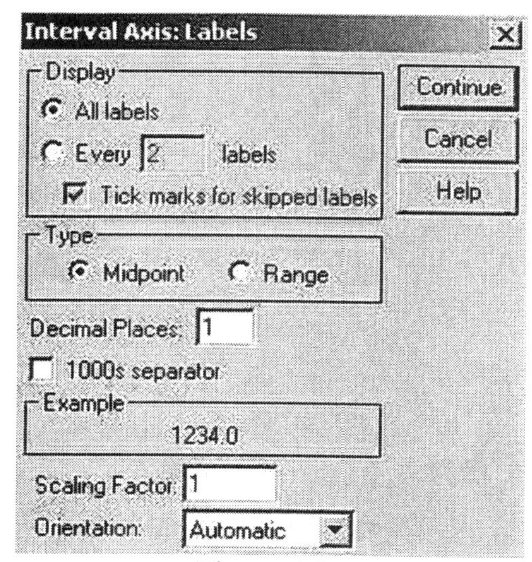

Figure 1.24

10. Click **Every ☐ labels** in the "Display" box. "Every 2 labels" is the default option.
11. Click **Tick marks for skipped labels**. The ✓ disappears.
12. Click **Range** within the "Type" box.
13. Within the "Decimal Places" box, change 1 to **0.**
14. Click ▼ in the "Orientation" box and then click **Horizontal.**
15. Click **Continue.**
16. Click **OK.**

To remove the descriptive statistics (Std. Dev., Mean, and N) in the legend, follow these steps:

1. Click **Chart** and then click **Legend.** The "Legend" window in Figure 1.25 appears.

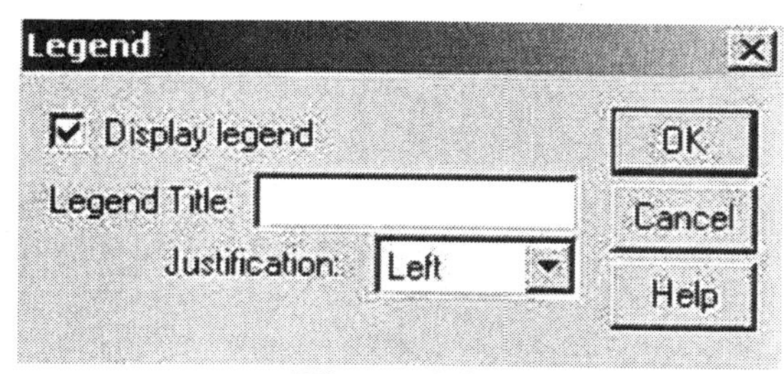

Figure 1.25

2. Click **Display legend.** The ✓ disappears.
3. Click **OK.**
4. If you are finished editing the chart, click **File** from the main menu bar within the "Chart – SPSS for Windows Chart Editor" window and then click **Close** to return to the "Output – SPSS for Windows Viewer" window.

Figure 1.26 is the resulting SPSS for Windows output (with the exception of the change in the *y* axis title). To add numbers to the bars, make changes to the *y* axis (such as adding an axis title and changing the spacing of the tick marks), or change the color or the fill of the bars, follow the directions given with Example 1.1 about editing bar charts.

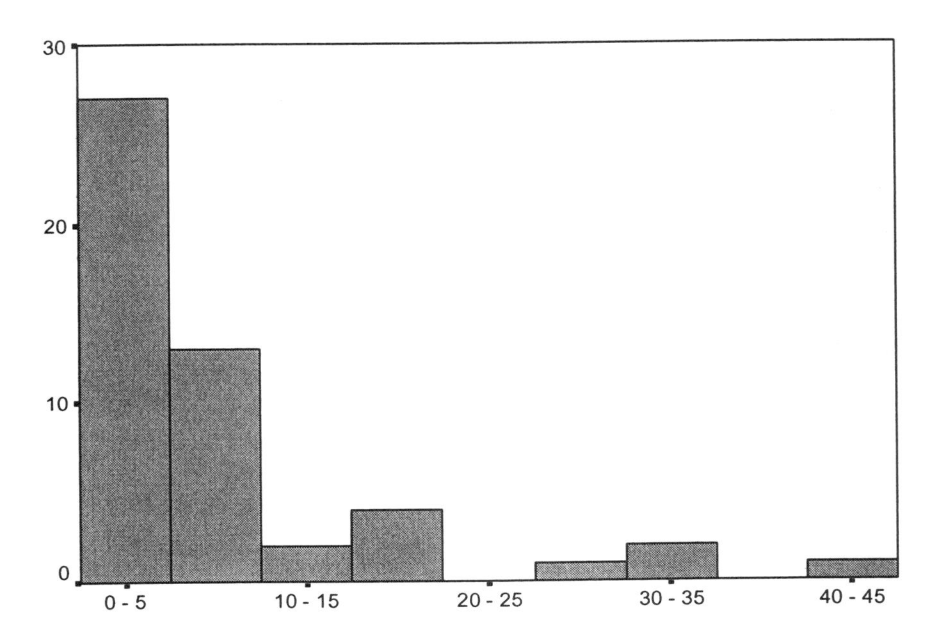

Figure 1.26

If you want a **relative frequency histogram** rather than a frequency histogram, follow the directions given earlier for editing bar charts changing the scale of the *y* axis from the counts (frequency) to the proportion of cases (relative frequency) for bar charts, for example.

The distribution shown in Figure 1.26 has a single peak, which represents states that are less than 5% Hispanic. The distribution is skewed to the right. The midpoint of the distribution is close to 4.7%. The spread is about 0% to 42%, but only four states fall above 20%.

Stemplots

A stemplot (also called a stem-and-leaf plot) gives a quick picture of the shape of the distribution for a quantitative variable while including the actual numerical values in the graph. Stemplots work best for small numbers of observations that are all greater than zero.

We will continue working with the data shown in Table 1.4 in BPS for the percentage of Hispanic residents in each of the 50 states in 2000. You might wish to reopen the file (***ta01-01.por***) if it is not still open. The SPSS for Windows Data Editor contains two variables called ***state*** (declared type string 30) and ***percent*** (declared type numeric 8.1).

To create a stemplot of this distribution, follow these steps:

1. Click **Analyze,** click **Descriptive Statistics,** and then click **Explore.** The "Explore" window in Figure 1.27 appears.

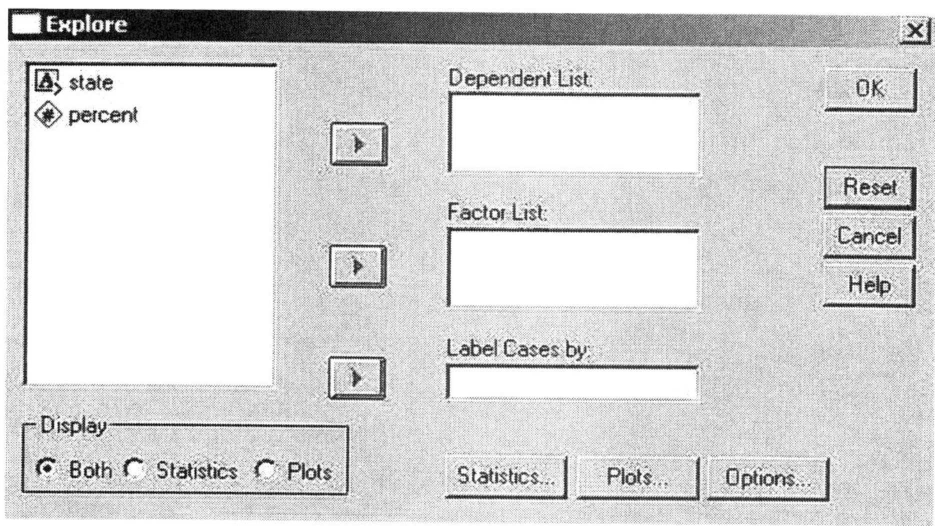

Figure 1.27

2. Click ***percent,*** then click ▸ to move ***percent*** into the "Dependent List" box.
3. By default, the "Display" box in the lower left corner has "Both" selected. Click **Plots** instead.
4. Next click **Plots** (located next to the "Options" button). The "Explore: Plots" window in Figure 1.28 appears.

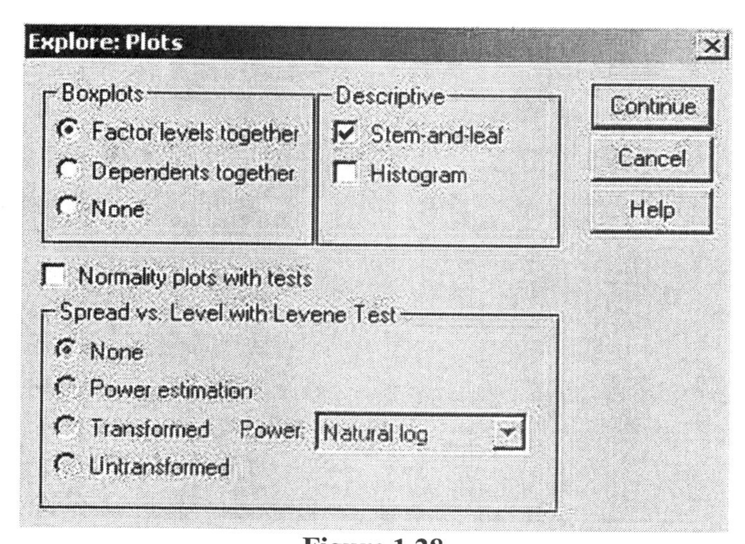

Figure 1.28

5. Click **None** within the "Boxplots" box. Be sure that a ✓ appears in front of "Stem-and-leaf" within the "Descriptive" box.
6. Click **Continue.**
7. Click **OK.**

Part of the resulting SPSS for Windows output is shown in Table 1.2.

```
PERCENT Stem-and-Leaf Plot

Frequency     Stem &  Leaf

10.00          0 .  0001111111
12.00          0 .  222222223333
 8.00          0 .  44444555
 6.00          0 .  667777
 4.00          0 .  8899
 1.00          1 .  0
 1.00          1 .  3
 1.00          1 .  5
 2.00          1 .  67
 5.00 Extremes     (>=20)

Stem width:     10.00
Each leaf:       1 case(s)
```

Table 1. 2

The stemplot was put together by using the whole-number part of the observation as the stem and the final digit (tenths) as the leaf.

Time plots

Many interesting data sets are time series, measurements of a variable taken at regular intervals over time. When data are collected over time, it is a good idea to plot the observations in time order. A time plot puts time on the horizontal scale of the plot and the quantitative variable of interest on the vertical scale.

Table 1.3 in BPS gives the average tuition and fees paid by college students at four-year colleges, both public and private, from the 1971- 1972 academic year to the 2001-2002 academic year. Make a time plot of these data. The data can be retrieved from the file called ***ta01-03.por.***

The SPSS for Windows Data Editor contains three variables called ***year*** (declared numeric 11.0), ***private*** (declared numeric 11.0), and ***public*** (declared numeric 11.0).

To create a time plot for this data set, follow these steps:

1. Click **Graphs** and then click **Sequence.** The "Sequence Charts" window shown in Figure 1.29 appears.
2. Highlight both ***private*** and ***public,*** then click ▸ to move them into the "Variables" box.
3. Click ***year*** and then click ▸ to move ***year*** to the "Time Axis Labels" box.
4. Click **OK.**

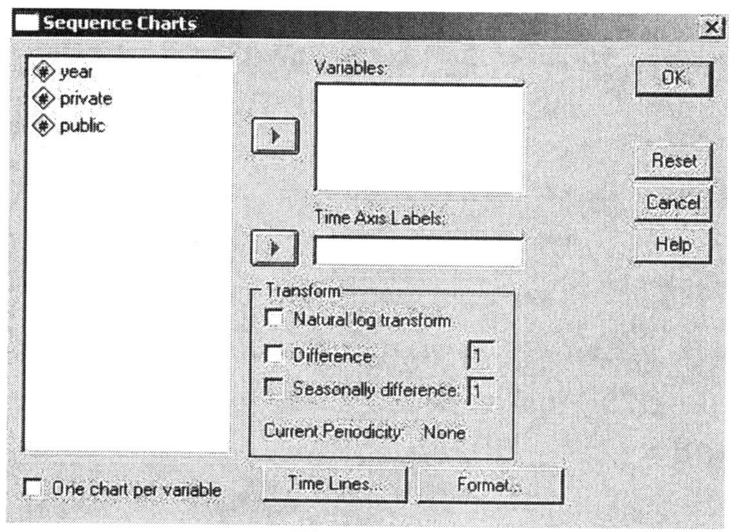

Figure 1.29

Figure 1.30 is the resulting SPSS for Windows output. To edit the x and y axes, consult the directions in the Editing Histograms section.

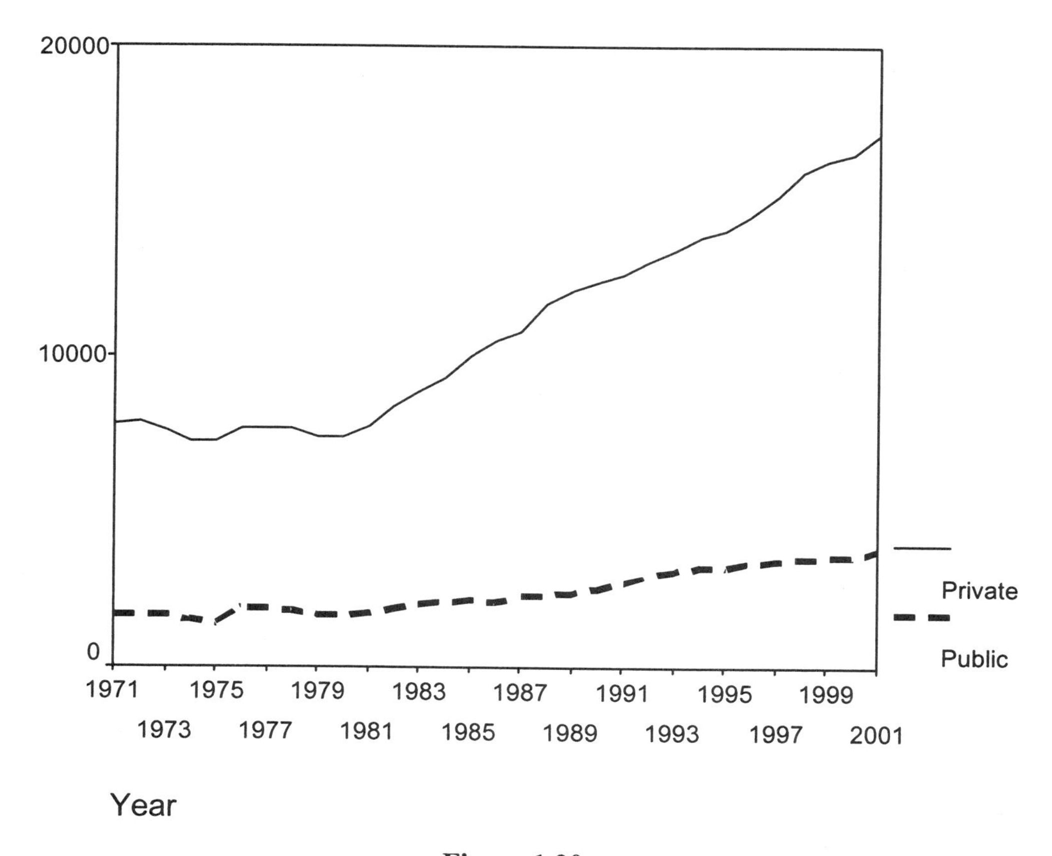

Figure 1.30

The time plot shows the steady increase in college tuition and fees for both private and public colleges from the 1971–1972 academic year to the 2001–2002 academic year.

Chapter 2. Describing Distributions with Numbers

Topics covered in this chapter:

- **Numerical Descriptions**
  - **Five-number summary**
  - **Center: mean, median, and mode**
  - **Spread: percentiles and standard deviation**
- **Visual Descriptions**
  - **Boxplots**
- **Comparing Distributions— Numerically & Visually**

This chapter introduces the notion of describing distributions with numbers. A description of a distribution almost always includes a **measure of center** and a **measure of spread.** To get a quick summary of both center and spread, use the five-number summary (the **minimum**, the **25th percentile**, the **median**, the **75th percentile,** and the **maximum**). The five-number summary leads to a visual representation of the distribution called a **boxplot.**

Numerical Descriptions

In the summer of 1998, Mark McGwire and Sammy Sosa captured the public's imagination with their pursuit of baseball's single-season home run record. McGwire eventually set a new standard with 70 home runs. How does this accomplishment fit McGwire's career? McGwire's home run counts for the year 1987 (his rookie year) to 2001 are given in BPS Exercise 1.26. Enter these data into the SPSS for Windows Data Editor using the variable name ***homeruns*** and declare it type numeric 8.0. Next, obtain descriptive statistics and a boxplot for the number of home runs (labeled ***homeruns*** in the following examples) that Mark McGwire hit in his first 15 major league seasons.

To obtain descriptive statistics (such as the mean, median, standard deviation, percentiles, etc.) for a quantitative variable (such as ***homeruns***), follow these steps:

1. Click **Analyze,** click **Descriptive Statistics,** and then click **Explore.** The "Explore" window (shown previously in Figure 1.34) appears, with the exception that the variable ***homeruns*** appears in the window.
2. Click ***homeruns,*** then click ▸ to move ***homeruns*** into the "Dependent List" box.
3. By default, the "Display" box in the lower left corner has **Both** selected. Click **Statistics.** *Note:* If you are also interested in obtaining a stemplot and a boxplot as part of the SPSS for Windows output at this time, then skip this step.
4. Click the **Statistics** button located in lower right corner of the window. The "Explore: Statistics" window in Figure 2.1 appears.

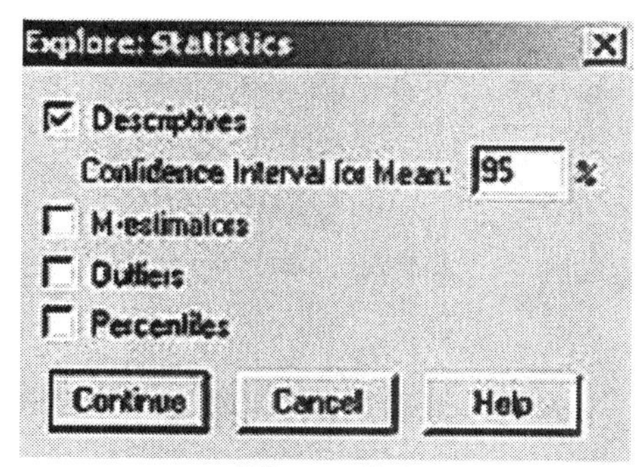

Figure 2.1

5. Click **Percentiles.** Be sure that a ✔ appears in front of "Descriptives."
6. Click **Continue.**
7. Click **OK.**

Tables 2.1 and 2.2 are part of the resulting SPSS for Windows output. They have been formatted as "Academic" instead of the default SPSS format.

Descriptives

			Statistic	Std. Error
Home Runs	Mean		38.6667	4.67890
	95% Confidence Interval for Mean	Lower Bound	28.6314	
		Upper Bound	48.7018	
	5% Trimmed Mean		38.5741	
	Median		39.0000	
	Variance		328.381	
	Std. Deviation		18.12128	
	Minimum		9.00	
	Maximum		70.00	
	Range		61.00	
	Interquartile Range		23.0000	
	Skewness		.017	.580
	Kurtosis		-.439	1.121

Table 2.1

Percentiles

		Percentiles						
		5	10	25	50	75	90	95
Weighted Average (Definition 1)	Home Runs	9.0000	9.0000	29.0000	39.0000	52.0000	67.0000	.
Tukey's Hinges	Home Runs			30.5000	39.0000	50.5000		

Table 2.2

The five-number summary consists of the minimum and maximum scores, the median, and the 1st and 3rd quartiles. Find the following numbers in Tables 2.1 and 2.2. The mean and the median number of home runs Mark McGwire hit in his first 15 major league seasons are 37.83 and 39, respectively. The number of home runs varies from a minimum of 9 to a maximum of 70. The standard deviation is 18.483.

The quartiles are shown in Table 2.2. SPSS for Windows uses slightly different rules than BPS to compute the quartiles, so the results given by the computer may not agree exactly with the results found by using BPS rules. SPSS for Windows uses two methods (labeled "Weighted Average [Definition 1]" and "Tukey's Hinges" [see Table 2.3]) to calculate the quartiles. The value for the 1st and 3rd quartiles differ slightly between the methods (24.5 versus 27 and 51.25 versus 50.5). In Table 2.2, they are labeled as the 25th and 75th percentiles, which has the same meaning as the 1st and 3rd quartiles.

Visual Descriptions

The five-number summary is also the basis for the boxplot shown below in Figure 2.4. To create a boxplot that is a visual rather than a numerical representation for a quantitative variable (such as ***homeruns*** from the McGwire example), follow these steps:

1. Click **Graphs** and then click **Boxplot.** The "Boxplot" window in Figure 2.2 appears.

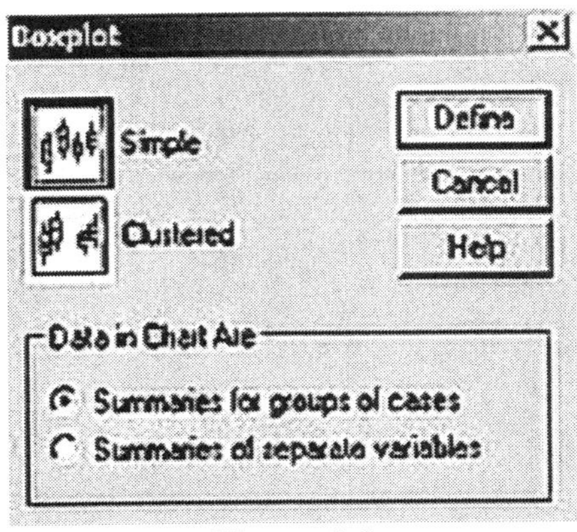

Figure 2.2

2. Click **Define.** The "Define Simple Boxplot: Summaries of Separate Variables" window in Figure 2.3 appears.

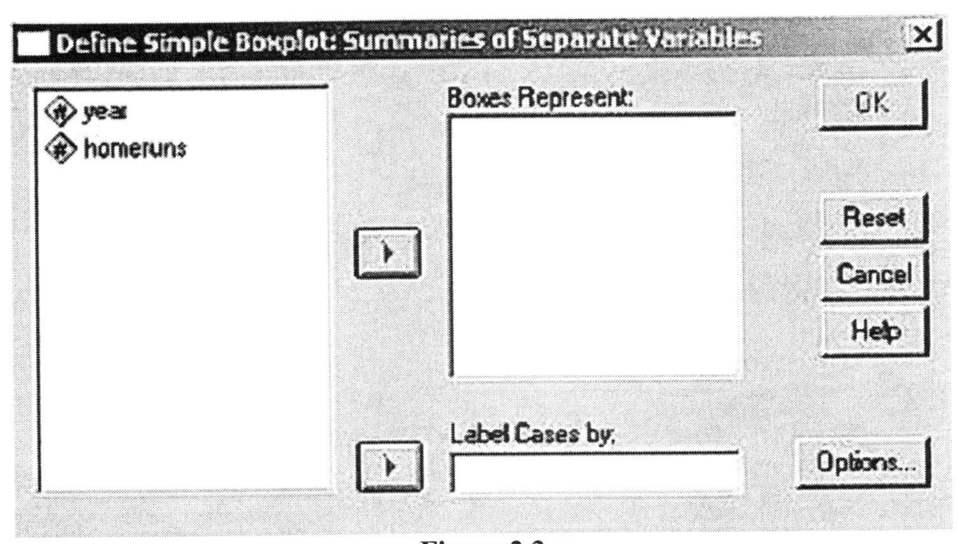

Figure 2.3

3. Click ***homeruns,*** then click ▸ to move ***homeruns*** into the "Boxes Represent" box.
4. Click **OK.**

Figure 2.4 is the resulting SPSS for Windows output. Notice that the horizontal lines in the boxplot mark the location of the minimum and maximum scores, the median, and the first and third quartiles. (To change the color or the pattern fill of the boxplot, follow the directions given in Chapter 1 about editing bar charts.)

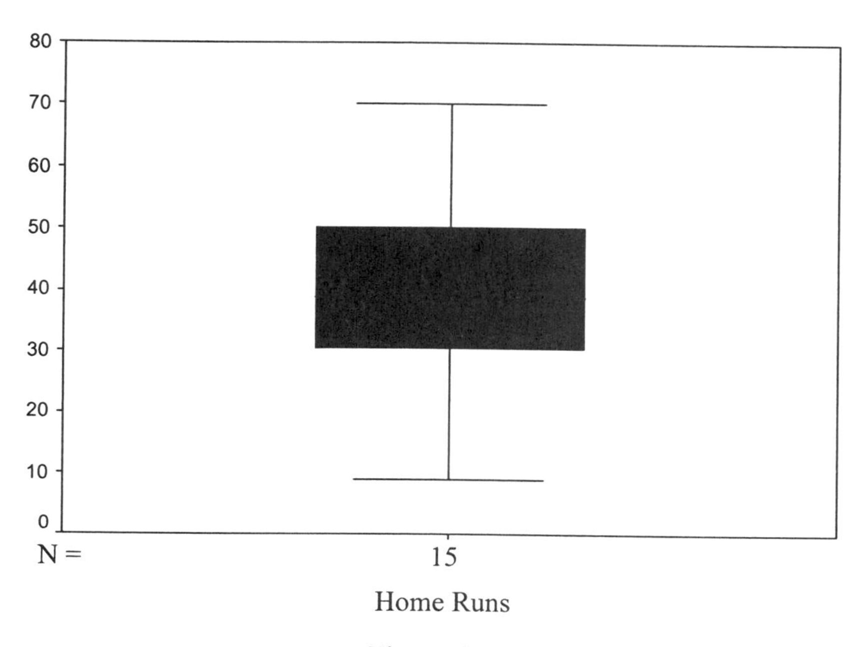

Figure 2.4

SPSS for Windows distinguishes between minor and major outliers (see Exercise 2.25 in BPS for a discussion of outliers). SPSS for Windows puts the observation (case) number beside the symbol used for the outlier. For the McGwire example, there are no outliers.

Comparing Distributions – Numerically & Visually

In this section, we will continue with the homerun data from BPS Chapter 1 Exercises 1.25 and 1.26. We might be interested in comparing McGwire's data with Babe Ruth's data. We can make the comparison by using both numerical and visual tools and by making side-by-side boxplots. First we need to add the Babe Ruth data to our SPSS Windows Data Editor and indicate which ***homeruns*** belong to which ***player.*** In the column adjacent to the ***homeruns,*** write "McGwire" beside each of the numbers already in the ***homeruns*** column. Define this variable ***player*** as type string 8. Next, enter Babe Ruth's home runs along with his name in these same two columns below the data for McGwire.

To obtain descriptive statistics (such as the mean, median, standard deviation, percentiles, etc.) for a quantitative variable broken down by a categorical variable (such as the home run data), follow these steps:

1. Click **Analyze,** click **Descriptive Statistics,** and then click **Explore.** The "Explore" window shown above in Figure 2.1 appears, with the appropriate variable names in the window.
2. Click ***homeruns,*** then click ▸ to move ***homeruns*** into the "Dependent List" box.

3. Click ***player,*** then click ▸ to move ***player*** into the "Factor List" box.
4. By default, the "Display" box in the lower left corner has **Both** selected. *Note:* If you are also interested in obtaining stemplots and side-by-side boxplots as part of the SPSS for Windows output then leave this button clicked and omit steps 5 and 6.
5. Click the **Statistics** button located in lower right corner of window. The "Explore: Statistics" window shown in Figure 2.1 appears.
6. Click **Percentiles.** Be sure that "Descriptives" is checked.
7. Click **Continue.**
8. Click **OK.**

Tables 2.3 and 2.4 are part of the resulting SPSS for Windows output.

Descriptives

	Player			Statistic	Std. Error
Home Runs	McGwire	Mean		38.6667	4.67890
		95% Confidence Interval for Mean	Lower Bound	28.6314	
			Upper Bound	48.7019	
		5% Trimmed Mean		38.5741	
		Median		39.0000	
		Variance		328.381	
		Std. Deviation		18.12128	
		Minimum		9.00	
		Maximum		70.00	
		Range		61.00	
		Interquartile Range		23.0000	
		Skewness		.017	.580
		Kurtosis		-.439	1.121
	Ruth	Mean		43.9333	2.90397
		95% Confidence Interval for Mean	Lower Bound	37.7049	
			Upper Bound	50.1617	
		5% Trimmed Mean		44.2593	
		Median		46.0000	
		Variance		126.495	
		Std. Deviation		11.24701	
		Minimum		22.00	
		Maximum		60.00	
		Range		38.00	
		Interquartile Range		19.0000	
		Skewness		-.538	.580
		Kurtosis		-.228	1.121

Table 2.3

Percentiles

			Percentiles						
		Player	5	10	25	50	75	90	95
Weighted Average (Definition 1)	Home Runs	McGwire	9.0000	9.0000	29.0000	39.0000	52.0000	67.0000	.
		Ruth	22.0000	23.8000	35.0000	46.0000	54.0000	59.4000	.
Tukey's Hinges	Home Runs	McGwire			30.5000	39.0000	50.5000		
		Ruth			38.0000	46.0000	51.5000		

Table 2.4

The mean and median number of home runs Mark McGwire hit in his first 21 major league seasons are 38.67 and 39, respectively, whereas the mean and median number of home runs Babe Ruth hit during his 15 seasons with the New York Yankees are 43.93 and 46, respectively. For McGwire, the number of home runs varies from 9 to 70 with a standard deviation of 18.12. For Ruth, the number of home runs varies from 22 to 60 with a standard deviation of 11.25.

To create side-by-side boxplots for a quantitative variable broken down by a categorical variable (such as for these home run data), follow these steps:

1. Click **Graphs** and then click **Boxplot.** The "Boxplot" window shown in Figure 2.2 appears.
2. Click **Summaries for groups of cases** and then click **Define.** The "Define Simple Boxplot: Summaries for Groups of Cases" window in Figure 2.5 appears.

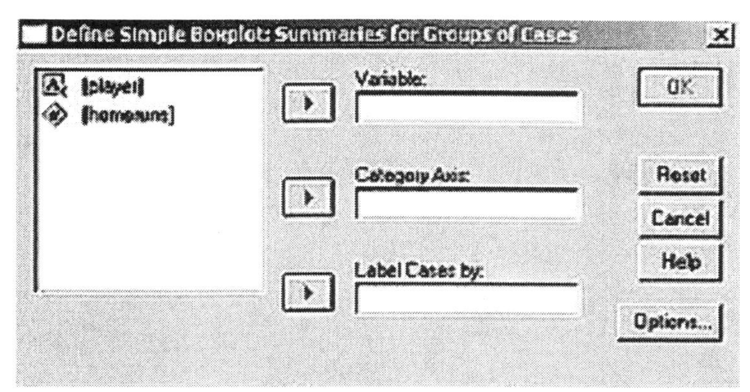

Figure 2.5

3. Click ***homeruns,*** then click ▸ to move ***homeruns*** into the "Variable" box.
4. Click ***player,*** then click ▸ to move ***player*** into the "Category Axis" box.
5. Click **OK.**

Figure 2.6 is the resulting SPSS for Windows output (except for the difference in the color scheme). To change the color or the pattern fill of the boxplots, follow the directions given in Chapter 1 about editing bar charts.

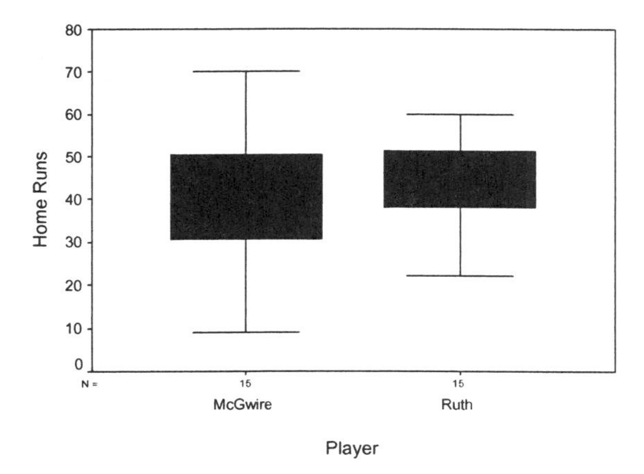

Figure 2.6

From the side-by-side boxplots, we see that Ruth is more consistent than McGwire; his home run counts are less spread out. His usual performance, as indicated by the median and spread (the "box" in the middle of the boxplot) of the middle half of the distribution, is a bit better than McGwire's.

Chapter 3. The Normal Distributions

Topics covered in this chapter:

- **Normal Distribution Probability Calculations**
- **Normal Quantile Plots**

One particularly important class of density curves is the **normal distribution.** These density curves are symmetric, unimodal, and bell-shaped. All normal distributions have the same overall shape. The exact density curve for a particular normal distribution is specified by giving its mean μ and its standard deviation σ. This section discusses how to use SPSS for Windows to do normal distribution probability calculations and how to obtain normal quantile plots.

Normal Distribution Probability Calculations

Using the Cumulative Distribution Function (CDF)

When we know information about a population including the variability (spread) and center (mean), we can use this information to answer questions about such variables as the IQ levels of second-year undergraduate students, the heights of women, or the number of golf strokes to complete a particular course. Recall that almost all variables take the shape of a normal curve and for the examples in this chapter we will assume that this is true for the variables at which we will look. What follows is an example about cholesterol levels in the blood of teenage boys. We will use SPSS to answer questions about the cholesterol levels of teenage boys based on what we know about the mean and standard deviation for cholesterol levels for this population.

The level of cholesterol in the blood is important because high cholesterol levels increase the risk of heart disease. The distribution of blood cholesterol levels in a large population of people of the same age and sex is roughly normal. For 14-year-old boys, the mean (μ) is 170 milligrams of cholesterol per deciliter of blood (mg/dl) and the standard deviation (σ) is 30 mg/dl. Using SPSS we will answer the following questions:

1. What proportion of 14-year-old boys have less than 240 mg/dl of cholesterol?
2. What proportion of 14-year-old boys have more than 240 mg/dl of cholesterol?
3. What proportion of 14-year-old boys have blood cholesterol between 170 and 240 mg/dl?

NOTES:
1. The choice of the numbers 170 and 240 is up to you as the scientist and are used here as illustrations.
2. In most cases, you can substitute the word percent (proportion * 100) for proportion, if that makes these calculations easier to understand.

To obtain the proportion of interest, follow these steps:

1. Enter the variables and values of ***x1*** and ***x2*** into the SPSS for Windows Data Editor, where ***x1*** = 170 (the mean [μ] for this population) and ***x2*** = 240 (a cholesterol level of interest) (see Figure 3.1).

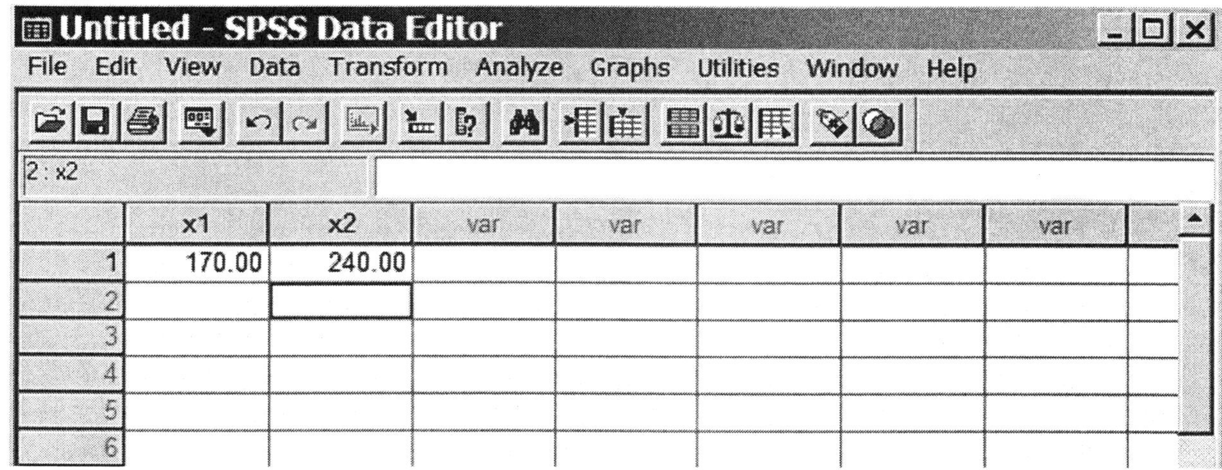

Figure 3.1

2. Click **Transform** and then click **Compute.** The "Compute Variable" window in Figure 3.2 appears.

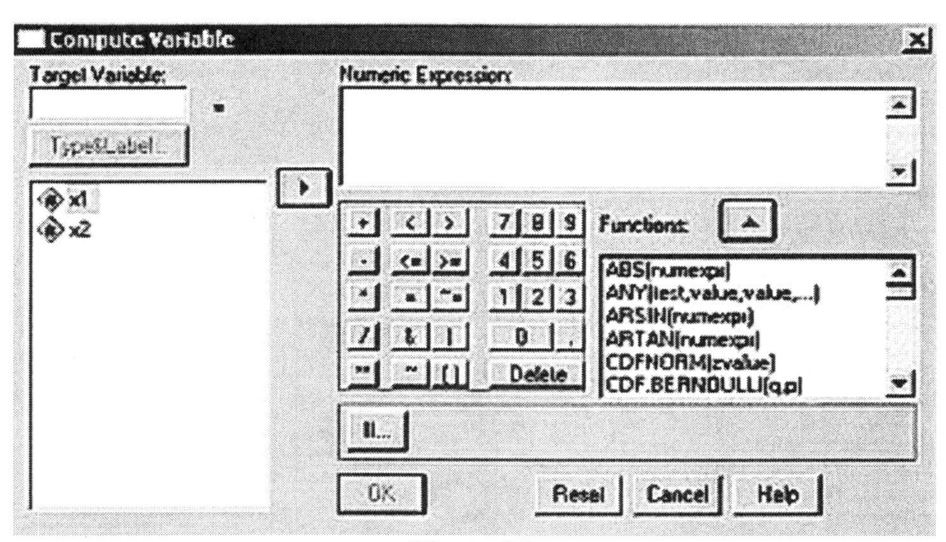

Figure 3.2

3. In the "Target Variable" box, type in ***prop.***
4. In the "Functions" box, click ▼ until **CDF.NORMAL(q, mean, stddev)** appears in the box. Double-click on **CDF.NORMAL(q, mean, stddev)** to move **CDF. NORMAL(?, ?, ?)** into the "Numeric Expression" box. The CDF.NORMAL(q, mean, stddev) function stands for the cumulative distribution function for the normal distribution, and it calculates the area to the left of q under the correct normal curve.
5. In the "Numeric Expression" box, change the second question mark to **170** and the third question mark to **30** (the appropriate values for the mean and the standard deviation).
6. For the proportion *less* than 240, **CDF.NORMAL(x2, 170, 30)** should appear in the "Numeric Expression" box (see Figure 3.3).

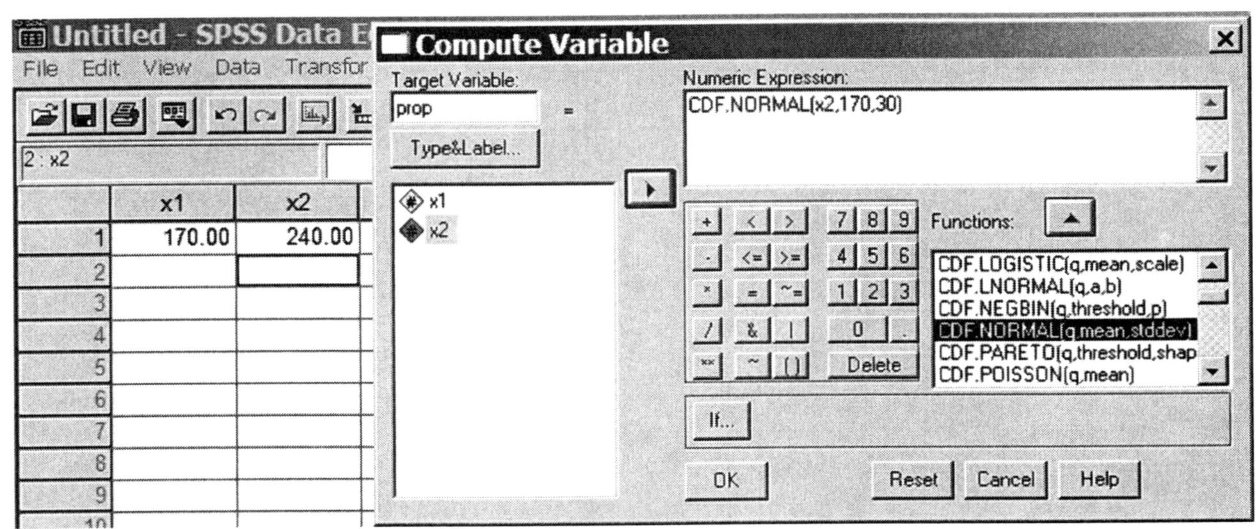

Figure 3.3

For the proportion *more* than 240, **1 – CDF.NORMAL(x2, 170, 30)** should appear in the "Numeric Expression" box.
For the proportion *between* 170 and 240, **CDF.NORMAL(x2, 170, 30) – CDF.NORMAL(x1, 170, 30)** should appear in the "Numeric Expression" box.

7. Click **OK.** This gives us the answer to our first question about blood cholesterol levels in 14-year-old boys.
8. Repeat this process for the remaining two questions (proportion *more* than 240, and proportion *between* 170 and 240). You may wish to change the "Target Variable" to ***prop2*** and then ***prop3*** so that you end up with all three values.

The variable ***prop*** can be found in the SPSS for Windows Data Editor. By default, the number of decimal places for the variable ***prop*** is two. The number of decimal places can be changed; follow the directions given in Chapter 0.

Here are the answers to the three questions posed above about blood cholesterol levels for teenage boys:

The proportion of 14-year-old boys with blood cholesterol less than 240 mg/dl is 0.9902 (or 99.02%). The proportion of 14-year-old boys with blood cholesterol levels higher than 240 mg/dl is 0.0098. The proportion of 14-year-olds with blood cholesterol between 170 and 240 mg/dl is 0.4902.

We were able to obtain these proportions because we knew the mean (μ) and standard deviation (σ) for the population of interest. It is these values that determine the middle and the spread of the normal curve that forms the basis for these calculations.

Using the Inverse Distribution Function (IDF) or "Backward" Normal Calculations

The previous example showed how to find the proportion of a given event. The next example reverses the process and shows how to find the observed value corresponding with a given proportion of observations above or below it. To complete these calculations, we need to know the population mean (μ) and standard deviation (σ) and the proportion of interest. The following example uses SAT scores. Because a large number or people write the scores each year, we know μ and σ.

Scores on the SAT verbal test in recent years follow approximately a normal distribution with a mean (μ) of 504 and a standard deviation (σ) of 111. Using SPSS, determine how high a student must score to place in the top 10% of all students taking the SAT.

After following the steps below, you can see that a student must score at least 646.25 to place in the top 10% of all students taking the SAT (90% of all students score at or below this SAT score).

To obtain the observed value corresponding to a given proportion, follow these steps:

1. Enter the variable and the value of ***x1*** into the SPSS for Windows Data Editor, where ***x1*** represents the area under the curve to the left of the desired score. In the SAT example, ***x1*** = 1 – 0.10 = 0.90. Click **Transform** and then click **Compute**. The "Compute Variable" window, as shown above in Figure 3.4, appears with ***x1*** in the window.
2. In the "Target Variable" box, type ***score.***
3. In the "Functions" box, click ▼ until **IDF.NORMAL(p, mean, stddev)** appears in the box. Double-click on **IDF. NORMAL(p, mean, stddev)** to move **IDF. NORMAL(?, ?, ?)** into the "Numeric Expression" box. The **IDF.NORMAL(p, mean, stddev)** function stands for the inverse of the cumulative distribution function for the normal distribution, and it calculates the X value such that the area to the left of X under the appropriate normal curve is *p*.
4. In the "Numeric Expression" box, change the first question mark to ***x1,*** the second question mark to **504** (μ), and the third question mark to **111** (σ). **IDF.NORMAL(x1, 504, 111)** should appear in the "Numeric Expression" box.
5. Click **OK.** The variable ***score*** can be found in the SPSS for Windows Data Editor (646.25).

Chapter 4. Scatterplots and Correlation

Topics covered in this chapter:

- **Scatterplots**
- **Adding Categorical Variables to Scatterplots**
- **Correlation**

This section introduces analysis of two variables that may have a linear relationship. In analyzing two quantitative variables, it is useful to display the data in a **scatterplot** and determine the **correlation** between the data. A scatterplot is a graph that puts one variable on the *x* axis and the other on the *y* axis, and it is used to determine whether an overall pattern exists between the variables. The correlation measures the strength and direction of the linear relationship, and the least-squares regression line is the equation of the line that best represents the data.

Scatterplots

Humans, in general, are interested in the patterns of nature. Scatterplots present a visual display of the way variables "go together" in the world. In the BPS Example 4.1, Ecologists take a look at the relationship between the size of a carnivore and how many of them there are in an area. The data can be found in the SPSS folder of your CD-ROM labeled ***ta04-01.por.***

To generate a scatterplot, follow these steps:.

1. Click **Graphs,** click **Scatter,** and click **Define.** The "Simple Scatterplot" window in Figure 4.1 appears.
2. Click ***abundance,*** then click ▸ to move ***abundance*** into the "Y Axis" box.
3. Click ***body mass,*** then click ▸ to move ***body mass*** into the "X Axis" box.
4. Click **OK.**

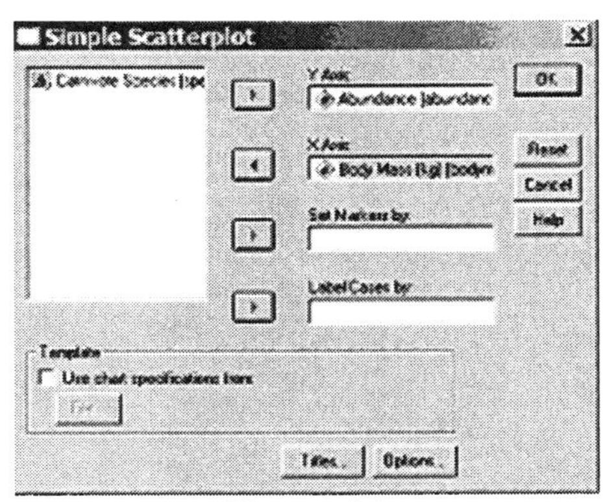

Figure 4.1

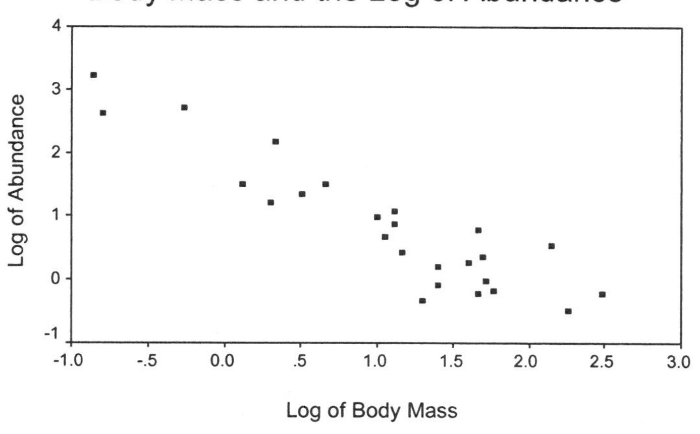

Figure 4.2

Figure 4.2 is a scatterplot of these data. Body mass is considered the explanatory variable and abundance the response variable. The scatterplot shows a strong, linear, negative relationship between body mass and abundance meaning that there are large numbers of small carnivores and few large ones.

Note that in the text, the data are measured on a logarithmic scale. The default in SPSS is linear therefore we must adjust the scales on both the *x*- and *y*-axes to make the graph look like the one in the text. This can be done using the **Transform ‣ Compute** options as shown in either Figure 4.3 or within the output page of SPSS. If using the "Compute Variable" window, choose the **Lg10 Function** and put your variable name inside the brackets. Add a new variable name in the "Target Variable:" window. The variable name used here is ***LogAbund.***

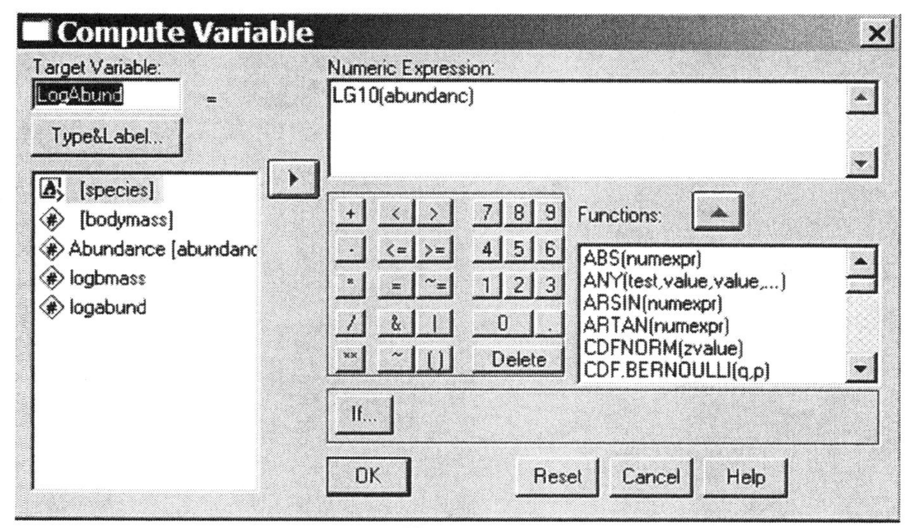

Figure 4.3

If instead you choose to edit the scale of the axes in the output page, when you have your graph selected (the black "handles" show around the edges), double-click on either the *x* or *y* axis. You will see a screen like the one in Figure 4.4. Click the button for logarithmic scale (linear is the default). Repeat this process for the other axis.

To change the type, size, and color of the symbols, double-click on the graph, then select the buttons for the color and symbol type. Using these options in the SPSS Chart Editor, adjust the chart to your specifications.

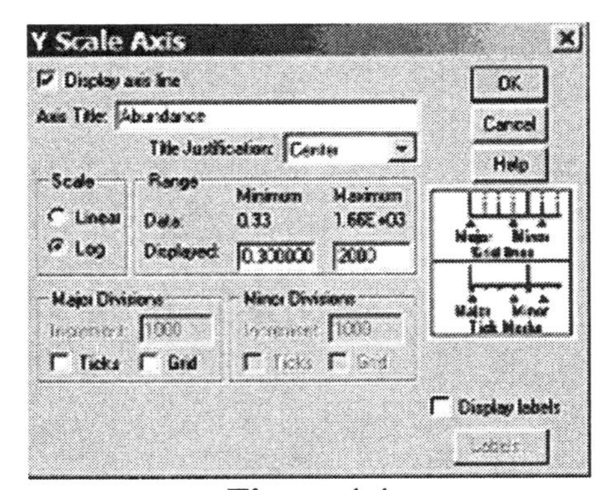

Figure 4.4

Adding Categorical Variables to Scatterplots

Sometimes it is useful to have some additional categorical information shown on the scatterplot. At the beginning of each semester I ask my students to anonymously provide their height, weight, shoe size, age, gender, and self-reported IQ. It is often useful to distinguish the data in a scatterplot based on a categorical variable. For example, when we correlate height and weight, adding gender may give us additional information in the scatterplot. Figure 4.5 shows the Simple Scatterplot Window for this example. Note that I have chosen to use Height rather than Weight as the explanatory variable.

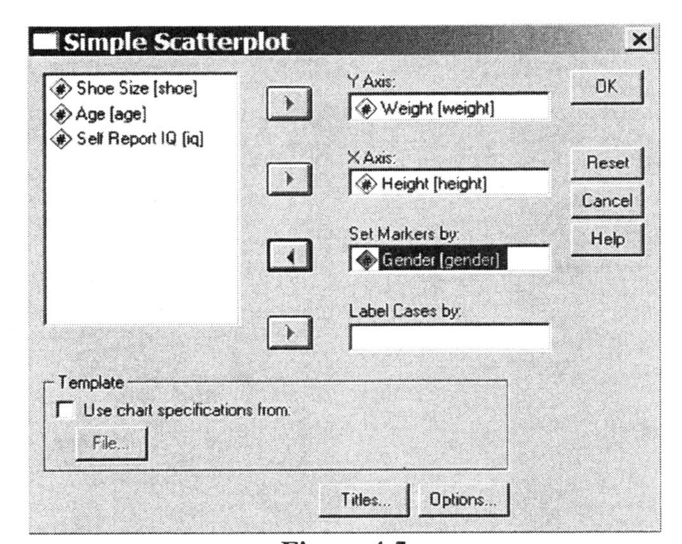

Figure 4.5

The scatterplot visualizing the relationship between Height and Weight for Males and Females is shown in Figure 4.6. The data for Males is shown by filled squares and for Females by open squares. As expected, males are generally taller than females and also generally weigh more. Thus, the data for Males tends to be in the upper right corner of the plot while the data for Females tends to be in the lower left quadrant. This appears to be a weak positive relationship.

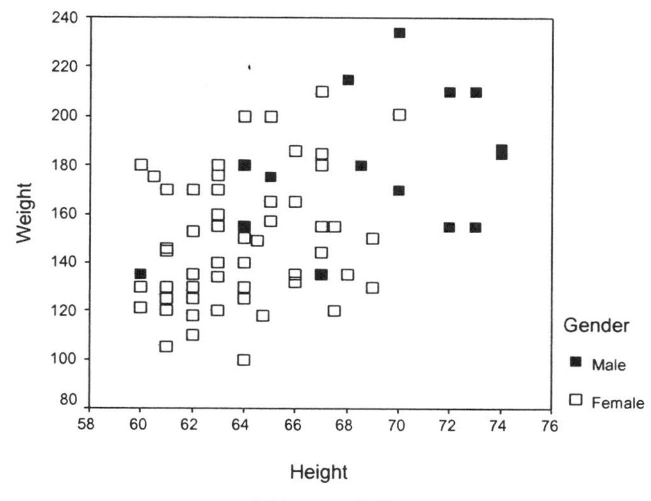

Figure 4.6

Correlation

While the scatterplot gives us visual information about the relationship between two quantitative variables, correlation gives us the numerical value for the relationship. If we return to the relationship between the abundance and body mass of carnivores (the data are located in ***ta04-01.por***), we can calculate the correlation between these two variables.

To obtain the correlation follow these steps:

1. Click **Analyze,** click **Correlate,** and then click **Bivariate,** as illustrated in Figure 4.7.

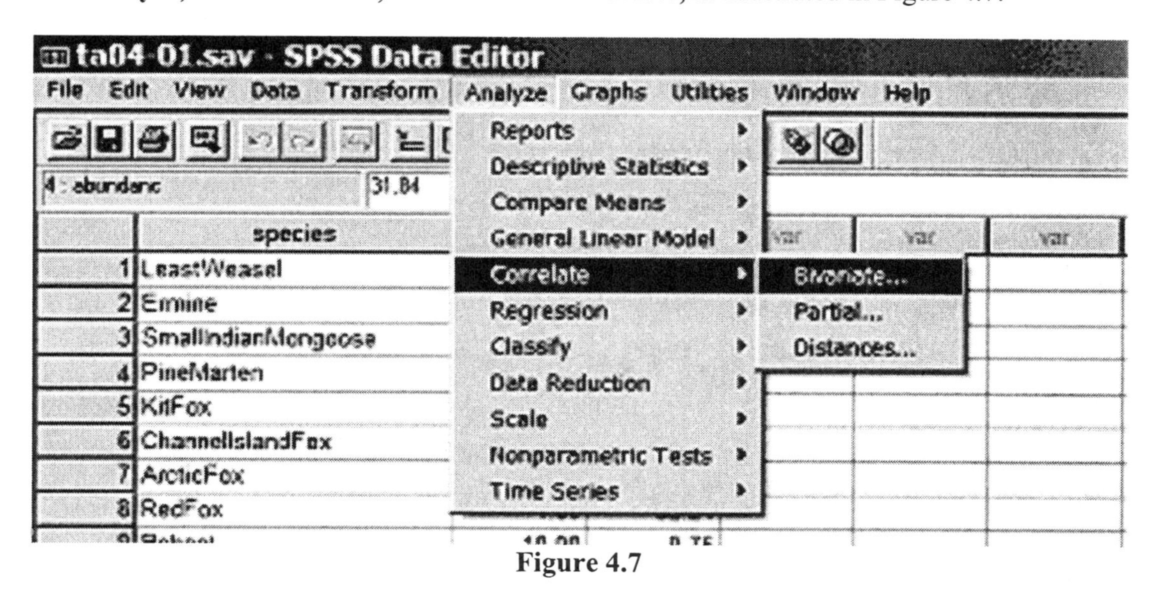

Figure 4.7

2. Click ***body mass,*** then click ‣ to move ***body mass*** into the "Variables" box.
3. Click ***abundance,*** then click ‣ to move ***abundance*** into the "Variables" box also. See Figure 4.8.
4. If you are interested in obtaining descriptive statistics for the variables, click the **Options** box, click **means and standard deviations,** then click **Continue.**

5. Click **OK.**

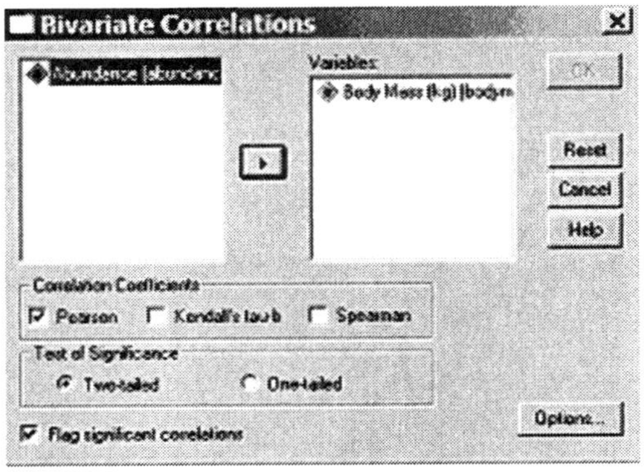

Figure 4.8

The correlation and the descriptive statistics for body mass and abundance are shown in Table 4.1. The descriptive statistics show the means and standard deviations for the original data. The correlation, as expected, is strongly negative at -.912. The correlation was calculated using the logarithm of the data rather than the original units.

Descriptive Statistics

	Mean	Std. Deviation	N
Body Mass	42.8728	70.72385	25
Abundance	115.2644	345.60913	25

Correlations

			Abundance
Body Mass	Pearson Correlation	1	-.205
	Sig. (2-tailed)	.	.324
	N	25	25
Abundance	Pearson Correlation	-.205	1
	Sig. (2-tailed)	.324	.
	N	25	25

Table 4.1

The descriptive statistics and correlation for the Height and Weight class data are shown in Table 4.2. These data are for males and females combined. The mean height for this sample is 64.86 inches and the mean weight is 153 pounds. The correlation between height and weight, as suggested by the scatterplot is positive but not very strong at .488. By the way, there is no correlation between IQ and shoe size, either!

Descriptive Statistics

	Mean	Std. Deviation	N
Height	64.86	3.515	79
Weight	153.60	28.940	78

Correlations

		Height	Weight
Height	Pearson Correlation	1	.488
	Sig. (2-tailed)	.	.000
	N	79	78
Weight	Pearson Correlation	.488	1
	Sig. (2-tailed)	.000	.
	N	78	78

Correlation is significant at the .01 level, 2-tailed significance.

Table 4.2

Chapter 5. Regression

Topics covered in this chapter:

- **Regression**
- **Fitted Line Plots**
- **Residuals**

Regression

The example used in this chapter is taken from BPS Exercise 4.4 and relates the count of new adult birds that join a bird colony to the percent of adult birds that return from the previous year. The data can be found in the SPSS data sets for Chapter 4 (***ex04-04.por***). The scatterplot for these data is shown in Figure 5.1 using percent returning (***Returning***) as the explanatory (x) variable and new adults (***NEW***) as the response (y) variable. Notice that there is a moderately strong, negative, linear correlation between these two variables. This scatterplot was produced using the instructions in Chapter 4.

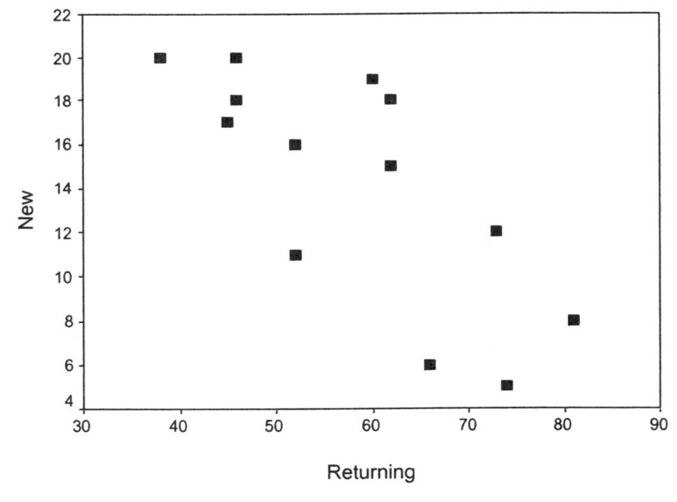

Figure 5.1

To obtain statistics such as the correlation coefficient, y-intercept, and slope of the regression line, follow these steps:

1. Click **Analyze,** ‣ **Regression,** ‣ **Linear.** The "Linear Regression" window in Figure 5.2 appears.
2. Click ***returning,*** then click ‣ to move ***returning*** into the "Independent(s)" box.
3. Click ***new,*** then click ‣ to move ***new*** into the "Dependent" box.
4. If you are interested in obtaining descriptive statistics for the variables, click **Statistics.** In the "Linear Regression Statistics" window, click **Descriptives,** then click **Continue.**
5. Click **OK.**

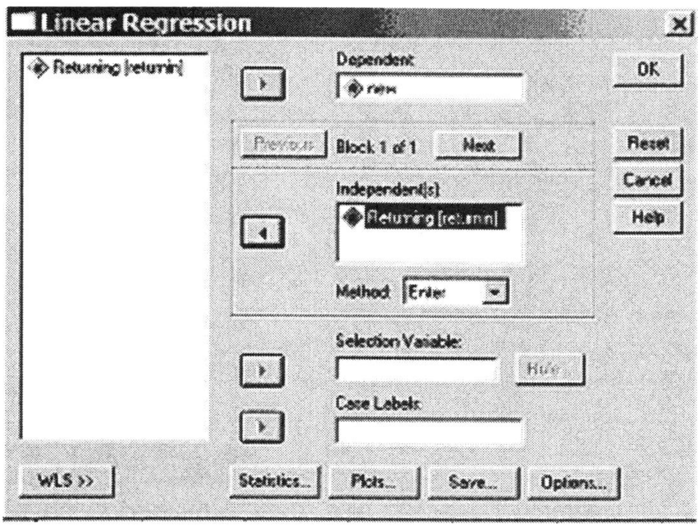

Figure 5.2

Tables 5.1 and 5.2 are part of the SPSS for Windows output that shows the correlation (**R**), r^2 (**R Square**), the y-intercept (labeled **Constant**), and the slope (labeled **Returning**).

Model Summary

Model	R	R Square	Adjusted R Square	Std. Error of the Estimate
1	.748[a]	.560	.520	3.667

a. Predictors: (Constant), Returning

Table 5.1

Coefficients[a]

Model		Unstandardized Coefficients		Standardized Coefficients	t	Sig.
		B	Std. Error	Beta		
1	(Constant)	31.934	4.838		6.601	.000
	Returning	-.304	.081	-.748	-3.743	.003

a. Dependent Variable: NEW

Table 5.2

From these tables, we can construct the regression equation as:

Predicted number of ***new*** = 31.934 + (-.304)(***Returning***)

These coefficients (values of the intercept (**Constant**) and slope (**Returning**)) are read from the "Unstandardized Coefficients" box in Table 5.2 and inserted in the appropriate places in the equation for a straight line (see Example 5.2 in BPS).

Fitted Line Plots

To plot the least-squares regression line on the scatterplot, follow these steps:

1. Generate the scatterplot as shown in Figure 5.1.
2. When the scatterplot appears in the Output window, double-click inside the scatterplot to gain access to the Chart Editor.
3. Once in the Chart Editor, click **Chart** ‣ **Options** ‣ **Total,** which is found in the "Fit Line" box of the "Scatterplot Options" window (see Figure 5.3).
4. Click **Fit Options.** The "Scatterplot Options: Fit Line" window in Figure 5.4 appears.
5. The "Fit Method" defaults to "Linear regression," which is the desired result.
6. If you are interested in displaying R^2 in the legend of the scatterplot, click **Display R-square in legend.**
7. Click **Continue.**
8. Click **OK.**

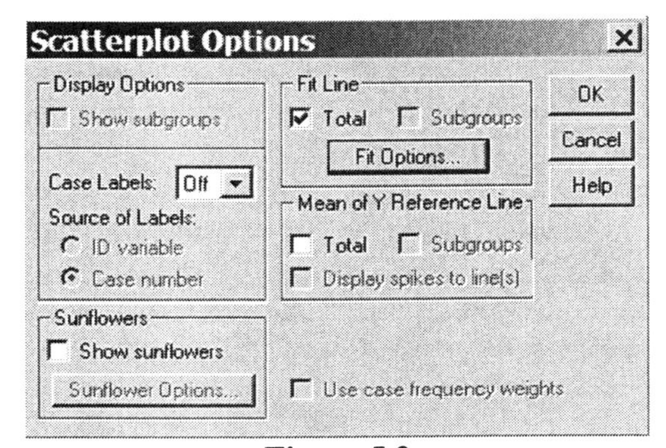

Figure 5.3

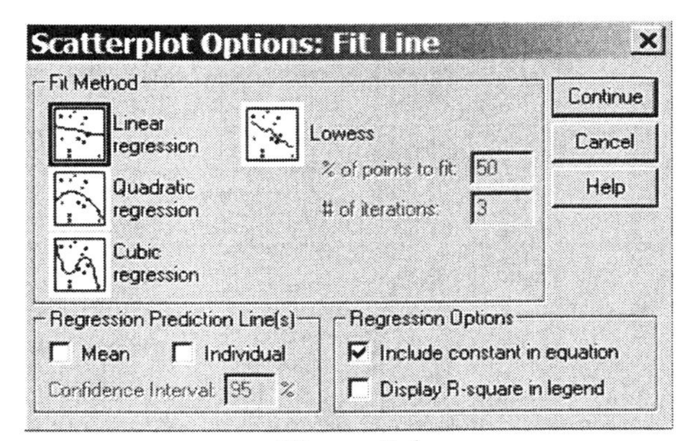

Figure 5.4

The resulting fitted line plot is shown in Figure 5.5. Compare this to Figure 5.1.

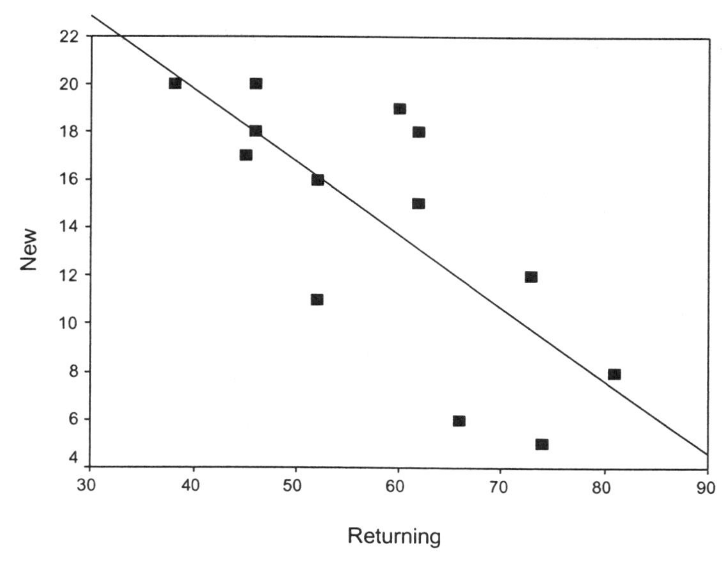

Figure 5.5

Residuals

A **residual** is the difference between an observed value of the response variable and the value predicted by the regression line, written $residual = y - \hat{y}$. The primary purpose for analyzing residuals is to determine whether or not the linear model best represents a data set.

Residual plots are a type of scatterplot in which the independent variable is generally on the x axis and the residuals are on the y axis. Such a plot helps us to assess the fit of a regression line.

It is desirable for no pattern to exist on the residual plot, that is for the plot to be an unstructured band of points centered around $y = 0$. If this is the case, then a linear fit is appropriate. If a pattern does exist on the residual plot, it could indicate that the relationship between y and x is nonlinear or that perhaps the variation of y is not constant as x increases. Residual plots are also useful in identifying outliers and influential observations.

To generate the residuals, follow these steps:

1. Click **Analyze** ‣ **Regression** ‣ **Linear.**
2. Click ***returning,*** then click ‣ to move ***returning*** into the "Independent(s)" box.
3. Click ***new,*** then click ‣ to move ***new*** into the "Dependent" box.
4. Click **Save,** then click **Unstandardized**, which is found in the "Residuals" box (see Figure 5.6).
5. Click **Continue,** then **OK.**

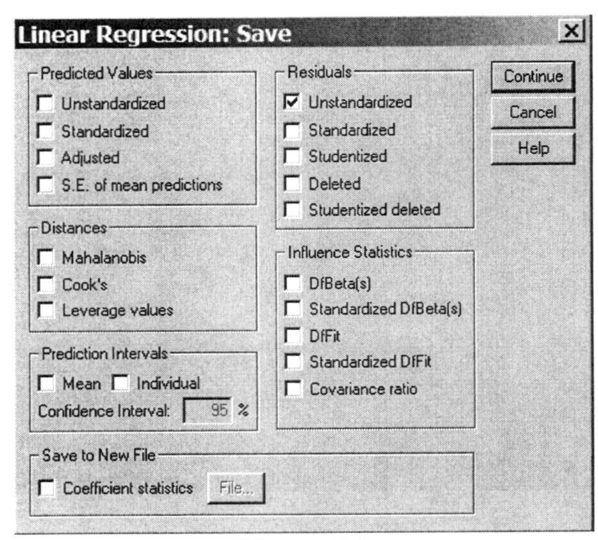

Figure 5.6

The residuals for this data set have been generated, saved, and added to the Data Editor, as shown in Figure 5.7.

Untitled - SPSS Data Editor

File Edit View Data Transform Analyze Graphs Utilities Window Help

0 : res_1

	returnin	new	res_1	var
1	74	5	-4.43656	
2	66	6	-5.86874	
3	81	8	.69160	
4	52	11	-5.12507	
5	73	12	2.25942	
6	62	15	1.91516	
7	52	16	-.12507	
8	45	17	-1.25323	
9	62	18	4.91516	
10	46	18	.05080	
11	60	19	5.30712	
12	46	20	2.05080	
13	38	20	-.38139	
14				
15				
16				

Data View / Variable View — SPSS Processor

Figure 5.7

To plot the residuals against the independent variable (***returning***), and to plot the reference line at $y = 0$ on this plot, follow these steps:

1. Click **Graphs** ‣ **Scatter** ‣ **Define.**
2. Click ***returning,*** then click ‣ to move ***returning*** into the "X Axis" box.
3. Click ***Unstandardized Residuals (res_1)***, then click ‣ to move ***Unstandardized Residuals (res_1)*** into the "Y Axis" box (see Figure 5.8).
4. Click **OK.**
5. When the scatterplot appears in the output window, double-click inside the scatterplot to gain access to the Chart Editor.

6. Once in the Chart Editor, click **Chart** ‣ **Options** ‣ **Total,** which is found in the "Fit Line" box of the "Scatterplot Options" window (recall Figure 5.2).
7. Click **Fit Options.** The "Scatterplot Options: Fit Line" window (as we saw in Figure 5.3) appears.
8. The "Fit Method" defaults to "Linear regression," which is the desired result.

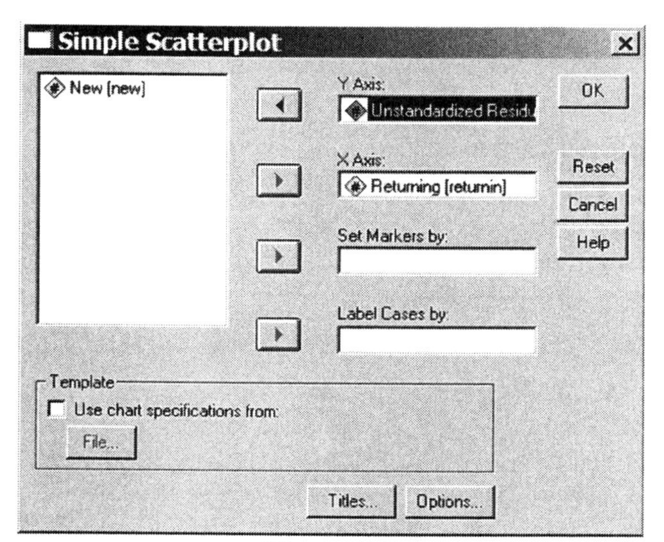

Figure 5.8

The residual plot, with the fit line at $y = 0$, is shown in Figure 5.9. Because there is no pattern to the points in this plot, the fit appears to be satisfactory.

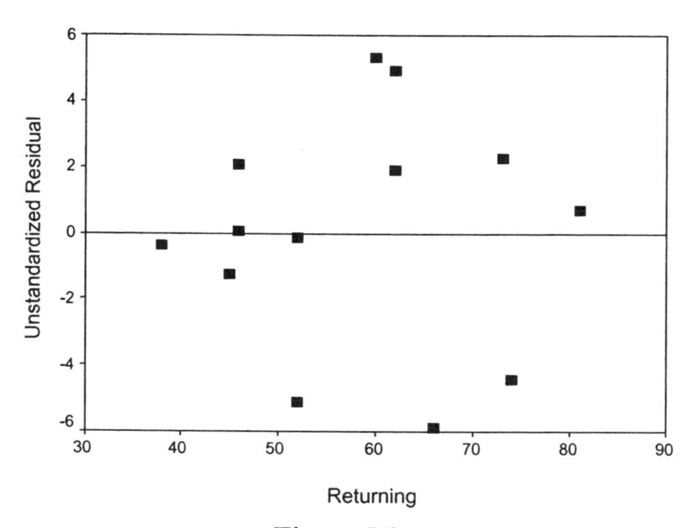

Figure 5.9

Chapter 6. Two-Way Tables

Topics covered in this chapter:

- **Two-Way Tables**
- **Row and Column Percents**
- **Graphing Row and Column Percents**
- **Three-Way Tables and Graphs**

Two-Way Tables

This chapter introduces the notion of analyzing two or more categorical variables using two-way and multi-way tables. The data used are from BPS Example 6.5 and in the SPSS Data Set for Chapter 6 (**eg06-05.por**). In this data set, we have information for about 1300 accident victims. For each victim, we have information about the patient's outcome (**outcome**), mode of transport (**transport**), and seriousness of the accident (**seriousn**). The data are in the form of string variables and are described on the "Variable View" tab in SPSS.

1. Click **Analyze** ‣ **Descriptive statistics** ‣ **Crosstabs.**
2. Click ***outcome,*** then click ‣ to move ***outcome*** into the "Row(s)" box.
3. Click ***transport,*** then click ‣ to move ***transport*** into the "Column(s)" box (see Figure 6.1).
4. Click **OK.**

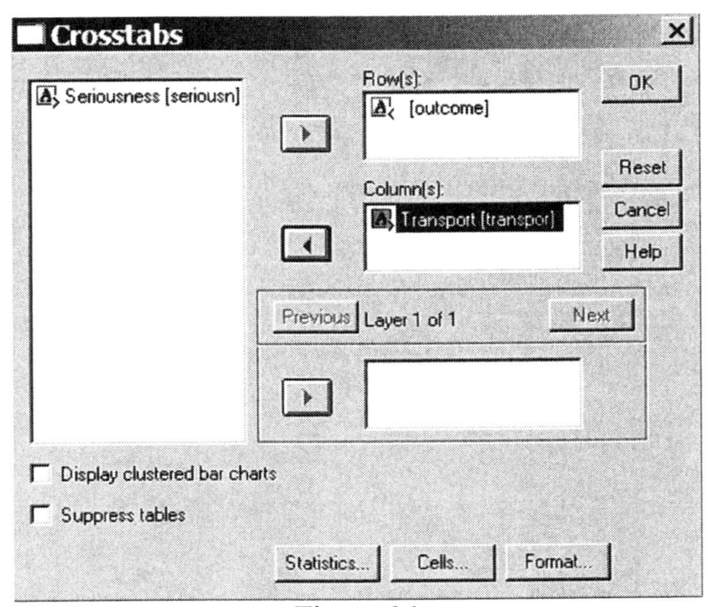

Figure 6.1

The resulting SPSS for Windows output is shown in Table 6.1.

OUTCOME * Transport Crosstabulation

Count

		Transport		Total
		Helicopter	Road	
OUTCOME	died	62	242	304
	survived	125	771	896
Total		187	1013	1200

Table 6.1

Note that we selected "Outcome" for the row variable and "Transport" for the column variable. In the "Crosstabs" and "Crosstabs: Cell Display" boxes in SPSS, you can change these choices to suit your interests. In Table 6.1, we see that 62/304 of victims transported by helicopter died compared to 242/304 victims transported by road.

Row and Column Percents

To describe the relationship between two categorical variables, you can generate and compare percentages. For example, we can determine how many patients survived each type of transport. Each cell count can be expressed as a percentage of the grand total, the row total, and the column total. To generate these percentages, follow these steps:

1. Click **Analyze** ‣ **Descriptive statistics** ‣ **Crosstabs.**
2. Click ***outcome,*** then click ‣ to move ***outcome*** into the "Row(s)" box.
3. Click ***transport,*** then click ‣ to move ***transport*** into the "Column(s)" box (see Figure 6.1).
4. Click the **Cells** button. The "Crosstabs: Cell Display" window in Figure 6.2 appears.
5. Click **Row, Column,** and **Total** within the "Percentages" box so a check mark (✔) appears before each type of percentage.
6. Click **Continue.**
7. Click **OK.**

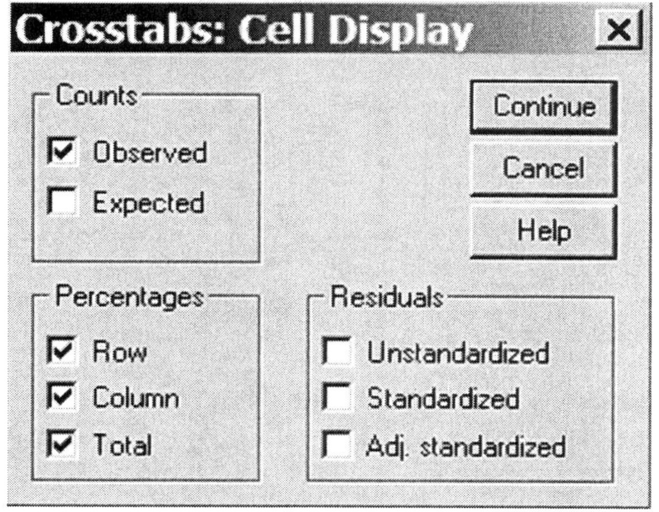

Figure 6.2

The outcome is shown in Table 6.2. This gives us the additional information that 33.2% of victims transported by helicopter died compared to 23.9% of victims transported by road.

OUTCOME * Transport Crosstabulation

			Transport		
			Helicopter	Road	Total
OUTCOME	died	Count	62	242	304
		% within OUTCOME	20.4%	79.6%	100.0%
		% within Transport	33.2%	23.9%	25.3%
		% of Total	5.2%	20.2%	25.3%
	survived	Count	125	771	896
		% within OUTCOME	14.0%	86.0%	100.0%
		% within Transport	66.8%	76.1%	74.7%
		% of Total	10.4%	64.3%	74.7%
Total		Count	187	1013	1200
		% within OUTCOME	15.6%	84.4%	100.0%
		% within Transport	100.0%	100.0%	100.0%
		% of Total	15.6%	84.4%	100.0%

Table 6.2

Graphing Row and Column Percents

To make a bar chart comparing the percent of victims who died for the two modes of transportation, follow these steps:

1. Click the **Graphs** ‣ **Bar** ‣ **Clustered** ‣ **Define** buttons. The "Define Clustered Bar: Summaries of Cases" window in Figure 6.3 appears.
2. Click "% of Cases" in the "Bars Represent" box.
3. Click ***outcome,*** then click ‣ to move ***outcome*** into the "Category Axis" box.
4. Click ***transport,*** then click ‣ to move ***transport*** into the "Define Clusters by" box.
5. Click **OK.**

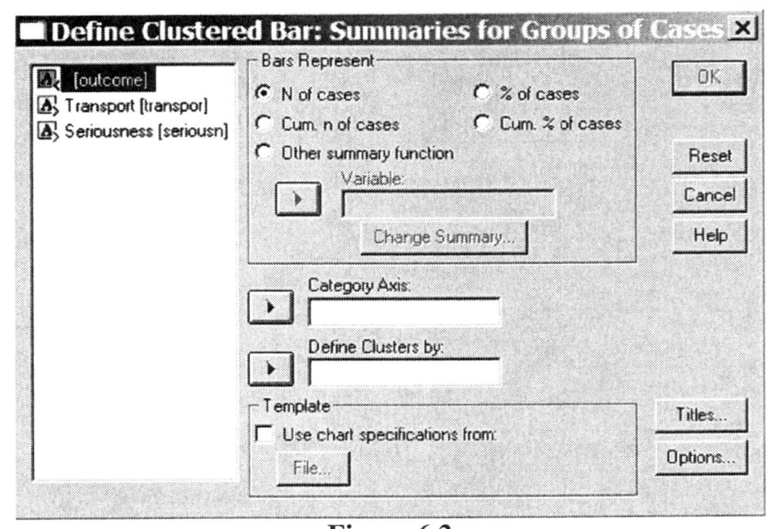

Figure 6.3

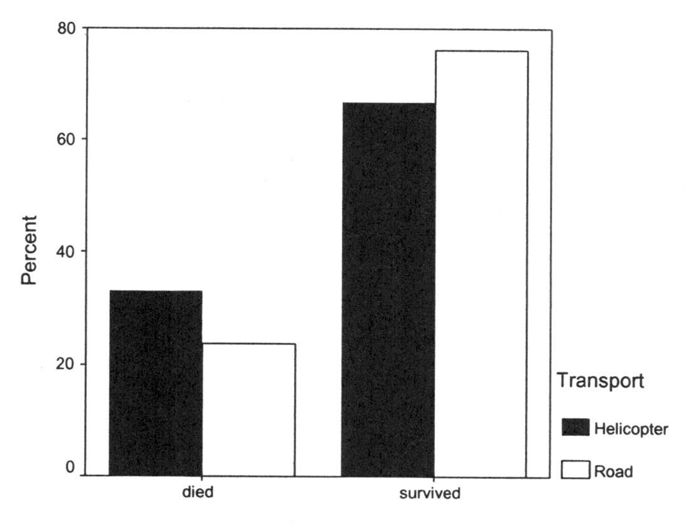

Figure 6.4

The resulting bar graph, shown in Figure 6.4, shows the same outcome as the numerical data: About 33% of the accident victims transported by helicopter died and about 22% of those transported by road died. This is probably because victims of more serious accidents tend to be transported more frequently by helicopter.

Three-Way Tables

To describe the relationship between three categorical variables, you can generate and compare percentages. For example, we can determine how many patients survived each type of transport and add information about the seriousness of the accident. To generate these percentages, follow these steps:

1. Click **Analyze** ‣ **Descriptive statistics** ‣ **Crosstabs.**
2. Click ***outcome,*** then click ‣ to move ***outcome*** into the "Row(s)" box.
3. Click ***transport,*** then click ‣ to move ***transport*** into the "Column(s)" box.
4. Click ***seriousn*** then click to move ***seriousn*** to the "Layer 1 of 1" box.
5. Click the **Display Clustered Bar Charts** button (see Figure 6.5).
6. Click the **Cells** button. The "Crosstabs: Cell Display" window in Figure 6.2 appears.
7. Click **Row, Column,** and **Total** within the "Percentages" box so a check mark (✔) appears before each type of percentage.
8. Click **Continue.**
9. Click **OK.**

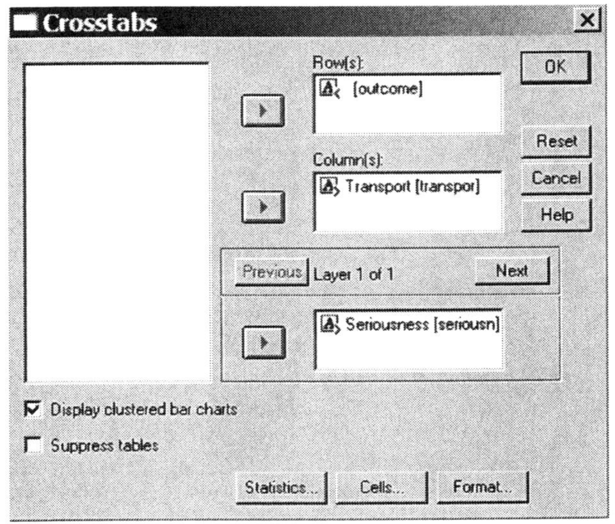

Figure 6.5

The SPSS output is shown below in Figures 6.6 and 6.7 along with Table 6.3. These are essentially two-way charts and tables for each value of seriousness (the third variable identified). When we examine these three-way tables, we can see that for less serious accidents, the survival rate is higher when victims are transported by helicopter. The same is true for more serious accidents.

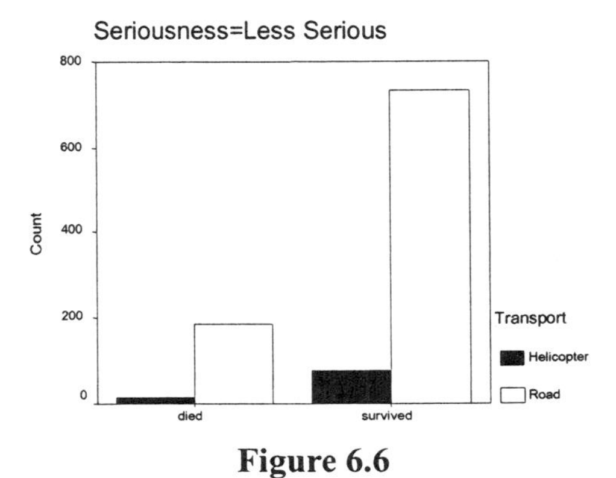

Figure 6.6

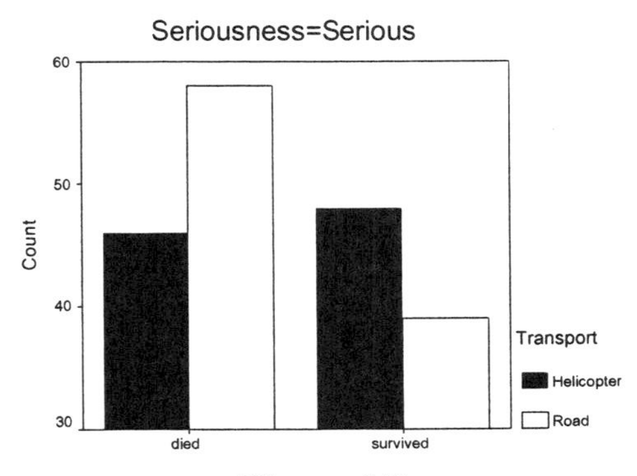

Figure 6.7

Case Processing Summary

	Cases					
	Valid		Missing		Total	
	N	Percent	N	Percent	N	Percent
* Transport * Seriousness	1200	100.0%	0	.0%	1200	100.0%

*** Transport * Seriousness Crosstabulation**

Seriousness			Transport		Total
			Helicopter	Road	
LessSerious	died	Count	16	184	200
		% within	8.0%	92.0%	100.0%
		% within Transport	17.2%	20.1%	19.8%
		% of Total	1.6%	18.2%	19.8%
	survived	Count	77	732	809
		% within	9.5%	90.5%	100.0%
		% within Transport	82.8%	79.9%	80.2%
		% of Total	7.6%	72.5%	80.2%
	Total	Count	93	916	1009
		% within	9.2%	90.8%	100.0%
		% within Transport	100.0%	100.0%	100.0%
		% of Total	9.2%	90.8%	100.0%
Serious	died	Count	46	58	104
		% within	44.2%	55.8%	100.0%
		% within Transport	48.9%	59.8%	54.5%
		% of Total	24.1%	30.4%	54.5%
	survived	Count	48	39	87
		% within	55.2%	44.8%	100.0%
		% within Transport	51.1%	40.2%	45.5%
		% of Total	25.1%	20.4%	45.5%
	Total	Count	94	97	191
		% within	49.2%	50.8%	100.0%
		% within Transport	100.0%	100.0%	100.0%
		% of Total	49.2%	50.8%	100.0%

Table 6.3

Chapter 7. Producing Data: Sampling

Topics covered in this chapter:

- **Random Samples**
- **Sorting Data**

This chapter demonstrates how SPSS for Windows can be used to **select *n* cases** from a finite population of interest using **simple random sampling**. Simple random sampling, the most basic sampling design, allows impersonal chance to choose the cases for inclusion in the sample, thus eliminating bias in the selection procedure.

Random Samples

BPS Example 7.6
Joan's small accounting firm serves 30 business clients, who are listed in Table 7.1. Joan wants to interview a sample of 5 clients in detail to find ways to improve client satisfaction. You can access these data from your CD-ROM by opening **eg07-06.por.**

01	A-1 Plumbing	09	Blue Print Specialties	17	Johnson Commodities	25	Rustic Boutique
02	Accent Printing	10	Central Tree Service	18	Keiser Construction	26	Satellite Services
03	Action Sport Shop	11	Classic Flowers	19	Liu's Chinese Restaurant	27	Scotch Wash
04	Anderson Construction	12	Computer Answers	20	MagicTan	28	Sewer's Center
05	Bailey Trucking	13	Darlene's Dolls	21	Peerless Machine	29	Tire Specialties
06	Balloons Inc.	14	Fleisch Realty	22	Photo Arts	30	Von's Video Store
07	Bennett Hardware	15	Hernandez Electronics	23	River City Books		
08	Best's Camera Shop	16	JL Records	24	Riverside Tavern		

Table 7.1

One option for choosing the five clients is to use a table of random digits. Another option is to use SPSS for Windows to randomly choose 5 of the 30 clients. A data file was created using ***client*** as a variable (declared as a string variable with length 24) that contained the names of the 30 clients.

To select a random sample of ***n*** cases from ***N*** cases, follow these steps:

1. Click **Data,** then click **Select Cases.** The "Select Cases" window in Figure 7.1 appears.

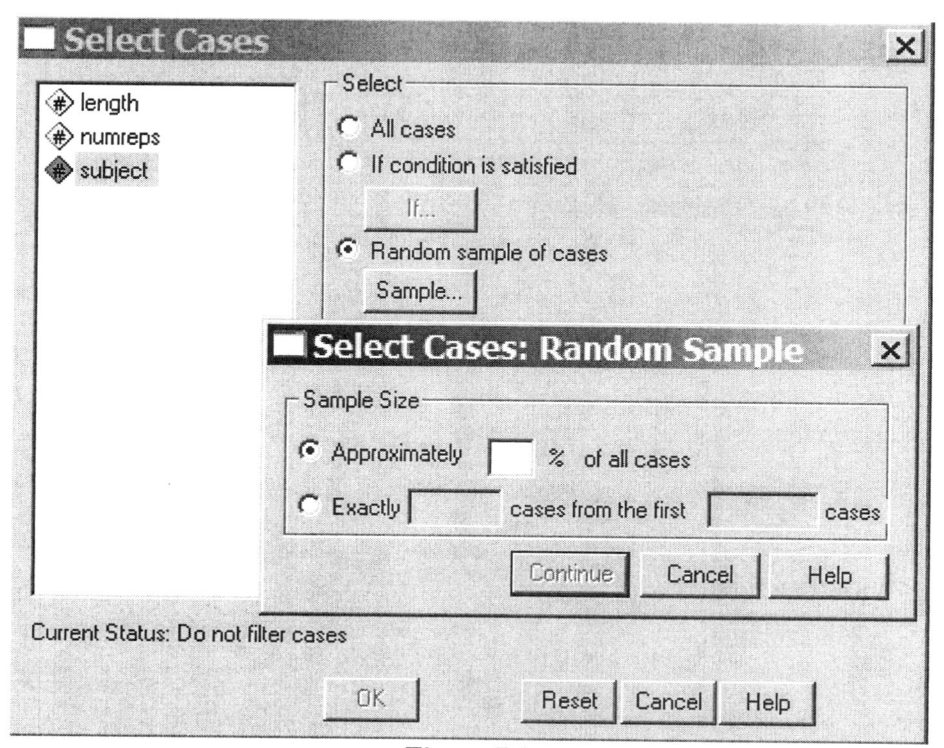

Figure 7.1

2. Click **Random sample of cases,** and then click the **Sample** button. The "Select Cases: Random Sample" window appears.
3. Click **Exactly ❍ cases from the first cases** and fill in the boxes so the line reads "Exactly **5** cases from the first **30** cases."
4. Click **Continue.**
5. Click **OK.**

SPSS for Windows creates a new variable, ***filter_$,*** which is assigned a value of 0 if the case was not randomly selected and a value of 1 if the case was randomly selected. Furthermore, unselected cases are marked in the Data Editor with an off-diagonal line through the row number, as can be seen in Figure 7.2, which shows the first 10 cases. With this particular procedure, clients A-1 Plumbing, Blue Print Specialties, and Central Tree Service are 3 of the 5 cases that were randomly selected to be interviewed. If the ***filter_$*** variable is deleted and the procedure repeated, you should obtain a different set of clients randomly selected to be interviewed.

	client	filter_$
1	A1Plumbing	1
2	AccentPrinting	0
3	ActionSportShop	0
4	AndersonConstruction	0
5	BaileyTrucking	0
6	BalloonsInc.	0
7	BennettHardware	0
8	Best'sCameraShop	0
9	BluePrintSpecialties	1
10	CentralTreeService	1

Figure 7.2

Sorting Data

The data in **eg07-06.por** are sorted in alphabetical order starting at "A." Suppose they were not sorted or that you wanted them in the reverse order. To sort data, follow these steps:

1. Click **Data,** then click **Sort Cases.** The "Sort Cases" window in Figure 7.3 appears.

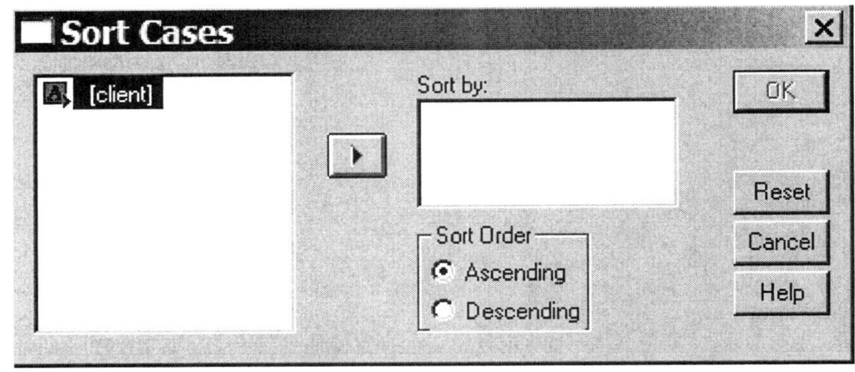

Figure 7.3

2. Click ***client,*** then click ▸ until ***client*** appears in the "Sort by:" box.
3. Click **descending.**
4. Click **OK.**

The data are now sorted from Von's Video Store to A1 Plumbing. A partial data set is shown in Figure 7.4.

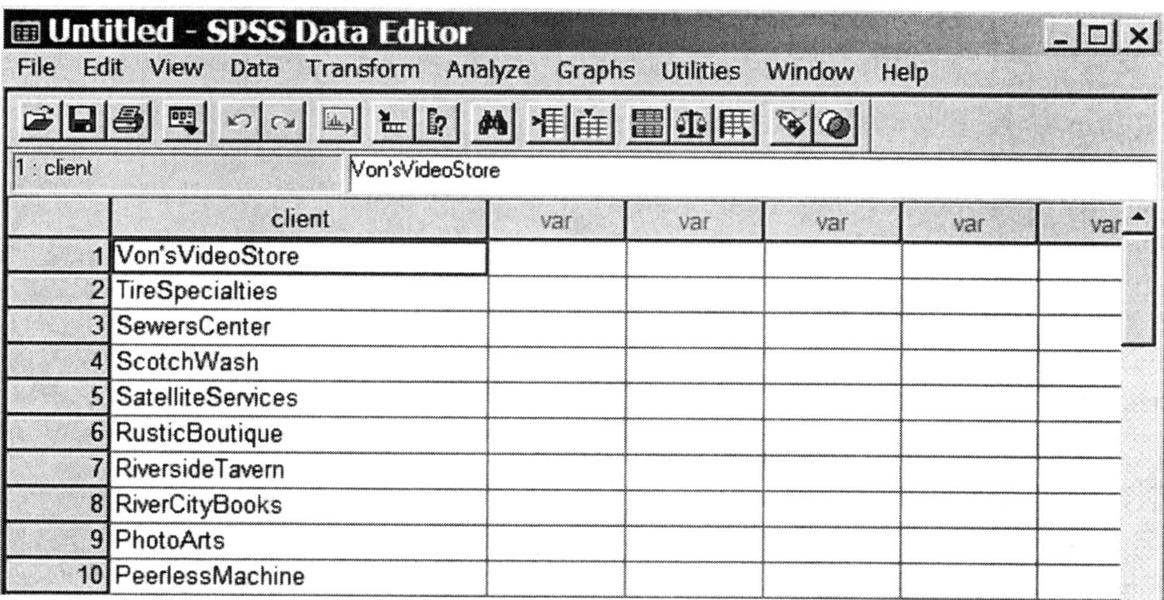
Untitled - SPSS Data Editor
File Edit View Data Transform Analyze Graphs Utilities Window Help
1 : client
Von'sVideoStore
client
var
var
var
var
var
1 Von'sVideoStore
2 TireSpecialties
3 SewersCenter
4 ScotchWash
5 SatelliteServices
6 RusticBoutique
7 RiversideTavern
8 RiverCityBooks
9 PhotoArts
10 PeerlessMachine

Figure 7.4

Chapter 8. Producing Data: Experiments

Topics to be covered in this chapter:

- **Randomization in Experiments**
- **Designs**

Randomization in Experiments

Sampling can be used to randomly select treatment groups in an experiment. In Example 8.4 in BPS, a food company assesses the nutritional quality of a new "instant breakfast" product by feeding it to newly weaned male white rats. The response variable is the rat's weight gain over a 28-day period. A control group of rats eats a standard diet but otherwise receives the same treatment as the experimental group. This experiment has one factor (the diet) with two levels. The researchers use 30 rats for the experiment and so must divide them into two groups of 15.

A data file was created using the variable ***rat*** (declared as numeric 8.0) that contained the numbers 1 to 30. To randomly assign the 30 rats to one of the two groups, use the following instructions. These are the same steps that were used in Chapter 7.

1. Click **Data,** then click **Select Cases.** The "Select Cases" window in appears.
2. Click **Random sample of cases,** and then click the **Sample** button.
3. Click **Exactly ❍ cases from the first cases,** and fill in the boxes so the line reads "Exactly **15** cases from the first **30** cases."
4. Click **Continue.**
5. Click **OK.**

The results for the first 10 cases (rats) are shown in Figure 8.1.

	rat	filter_$
1	1.00	1
2	2.00	1
3	3.00	0
4	4.00	1
5	5.00	1
6	6.00	0
7	7.00	1
8	8.00	0
9	9.00	1
10	10.00	1

Figure 8.1

Rats 1, 2, 4, 5, 7, 9, 10, 15, 17, 24, 26, 27, 28, 29, and 30 were randomly selected to receive one treatment (the experimental group). The remaining 15 rats would then receive the other treatment (the control group). If the ***filter_$*** variable is deleted and the procedure repeated, a different random assignment of rats to the two levels should be obtained.

Designs

Sampling can also be used to select treatment groups for more complex experimental designs. In Example 8.2 of BPS, the experimenters investigated the effects of repeated exposure to an advertising message. In this experiment, all subjects will watch a 40-minute television program that includes ads for a digital camera. Subjects will see either a 30-second or 90-second commercial repeated one, three, or five times. This experiment has two factors: length of commercial (with two levels) and number of repetitions (with three levels). The six different combinations of one level of each factor form six treatments.

First, enter the experimental design in an SPSS for Windows Data Editor by listing all the possible combinations and an identifying number for each subject. It requires at least 18 subjects to complete this experiment. The design is shown in Figure 8.2.

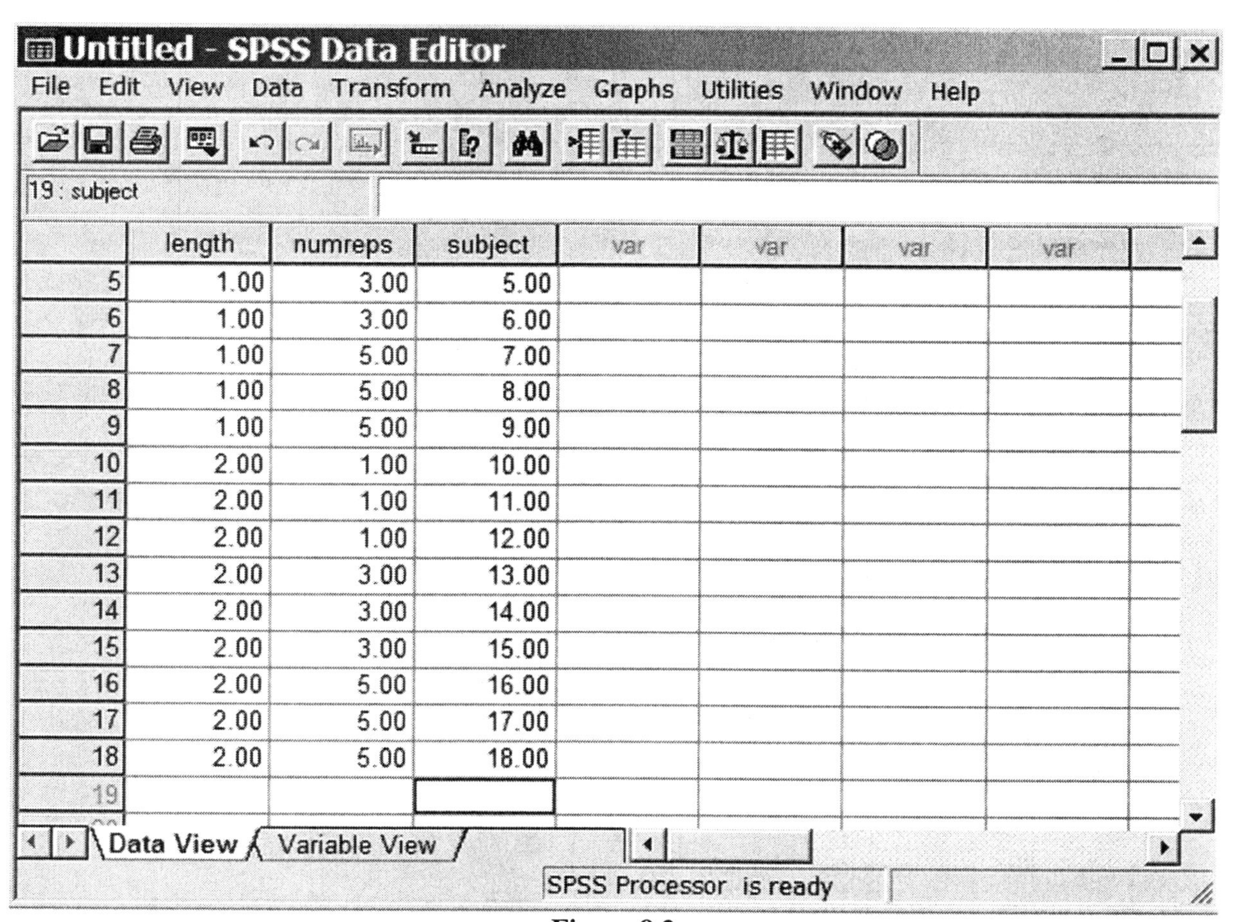

	length	numreps	subject	var	var	var	var
5	1.00	3.00	5.00				
6	1.00	3.00	6.00				
7	1.00	5.00	7.00				
8	1.00	5.00	8.00				
9	1.00	5.00	9.00				
10	2.00	1.00	10.00				
11	2.00	1.00	11.00				
12	2.00	1.00	12.00				
13	2.00	3.00	13.00				
14	2.00	3.00	14.00				
15	2.00	3.00	15.00				
16	2.00	5.00	16.00				
17	2.00	5.00	17.00				
18	2.00	5.00	18.00				
19							

Figure 8.2

To randomly assign subjects to treatments, follow the instructions shown above for randomization in experiments.

Chapter 9. Introducing Probability

Topics covered in this chapter:

- **Simulating Random Data**
- **Summarizing the Results**
- **Simulating Other Distributions**

Simulating Random Data

SPSS can be used to simulate random data. Simulating a series of coin tosses will be used as a starting example. To generate a series of 0's and 1's and then to recode them to Heads (H) and Tails (T), follow these steps:

1. Enter a number of your choosing in a new numeric variable in an SPSS for Windows Data Editor column. Repeat that same number 15 times. The number 125 was chosen for this example.
2. Next choose **Transform** ▸ **Compute** ▸ and the "Compute Variable" window appears.
3. Scroll down the "Functions:" box until ▸ **RV.Bernoulli(p)** appears. Click on this option to move **RV.Bernoulli(?)** into the "Numeric Expresion:" box.
4. Click on the **"?"** and change it to **.5** (for equal probabilities of Heads or Tails). Type a variable name in the "Target Variable:" box (see Figure 9.1).

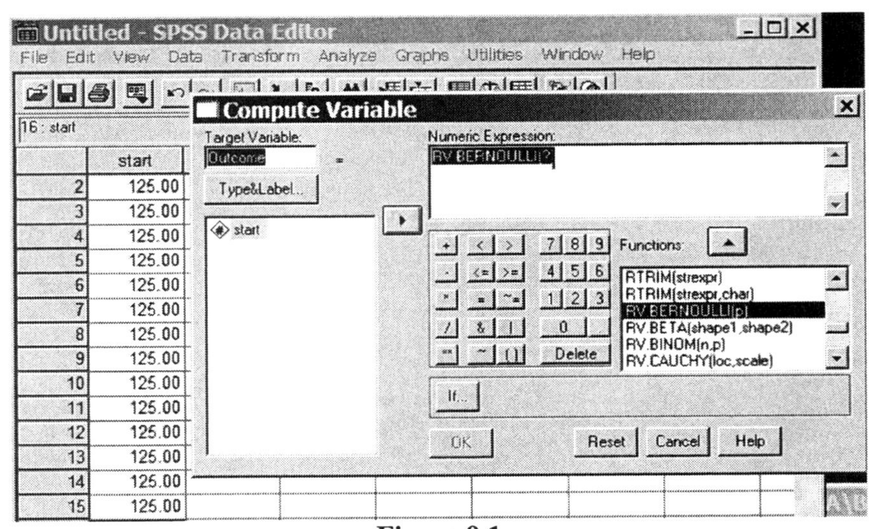

Figure 9.1

5. Click **OK.**
6. Now Click **Transform** ▸ **Recode** ▸ **Into Different Variable.**
7. Click ***outcome,*** then click ▸ to move ***outcome*** into the "Numeric Variable → Output Variable Box."
8. In the "Output Variable" box, type the name for your new variable, and in the "Label:" box expand on the name if you would like. Click **Change** (see Figure 9.2).

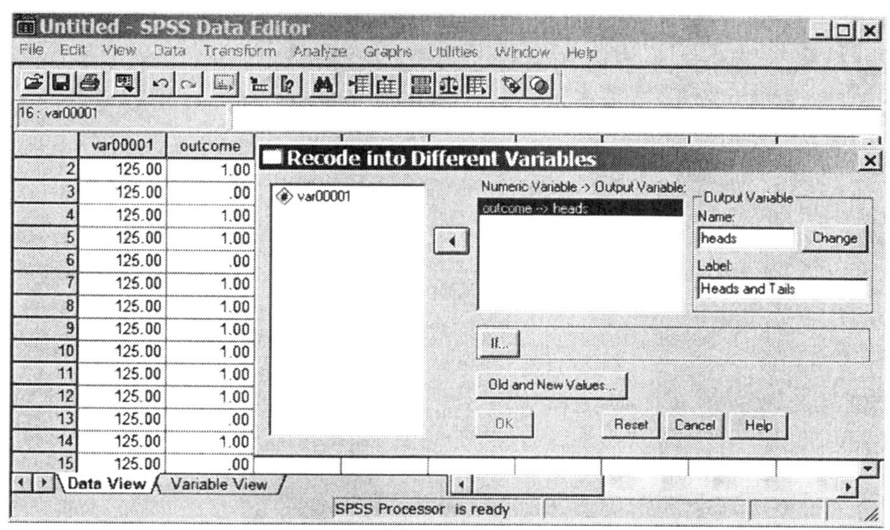

Figure 9.2

9. Click **Old and New Values.**
10. The "Recode into Different Variables: Old and New Values" box will appear. Click on the **"Output** Variables are Strings" box in the lower right corner.
11. In the "Old Value" box type ***1.*** In the "New Value" box, type ***H*** then click **Add. 1 → 'H'** appears in the "Old → New box."
12. Repeat this process to recode **0** to **T.**
13. Click **Continue.**
14. Click **OK.**

The results are shown in Figure 9.3. Note that because this is a random generation process, your list of **1's** and **0's** (and therefore your list of **H's** and **T's**) will likely be different from the one shown in the figure.

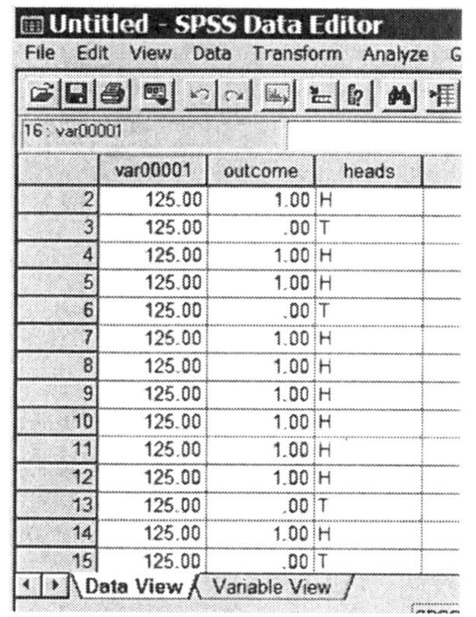

Figure 9.3.

Summarizing the Results

To tabulate the results of your 15 coin tosses, follow these steps:

1. Click **Analyze** ‣ **Descriptives** ‣ **Frequencies.**

2. Click ***Heads and Tails*** and ▸ to move ***Heads and Tails*** into the "Variable(s):" box.
3. Click **Charts** ▸ **Bar Charts** ▸ **Continue** (if you wish a visual description of your outcome in addition to the numerical one).
4. Click **OK.**

The outcome is shown below in Figure 9.4 and Table 9.1.

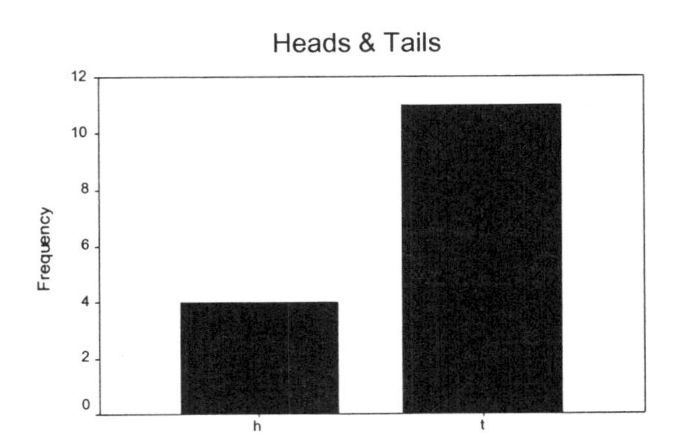

Figure 9.4

Statistics

Heads and Tails

N	Valid	15
	Missing	0

Heads and Tails

		Frequency	Percent	Valid Percent	Cumulative Percent
Valid	H	11	73.3	73.3	73.3
	T	4	26.7	26.7	100.0
	Total	15	100.0	100.0	

Table 9.1

As you look at this output, you may be surprised to see that the number of Heads is greater than the number of Tails. Recall that these will be equal only at .5 *in the long run*. To get a more realistic outcome, you need to toss the coin thousands of times. You can approximate such an outcome by repeating the steps thus far in this chapter after copying your number (125.00 in my example) and pasting it in starting at Row 16 until you have many repetitions of your coin toss. The more repetitions you have, the more your outcome will look like a binomial distribution.

Simulating Other Distributions

In most universities, the final grades in courses that are fairly large are normally distributed with a mean of 50 and a standard deviation of 15. Such a distribution makes the Dean and Registrar quite happy. Any grade distribution that is skewed toward either the high or the low end will be a cause for concern. The definition of a large class is not clearly specified but a class of 50 has been chosen for this example.

To generate several normally distributed random samples for classes of size 50 with a mean of 50 and a standard deviation of 15, follow these steps:

1. First, we need a "starting" number. In the first column of your data set, type in a number of your choosing and repeat the number until it occurs 50 times in that column. The number 52 has been used in this example and named as ***start.***
2. Click **Transform** ▸ **Compute** ▸ and the "Compute Variable" window appears.
3. Scroll down the "Functions:" box until ▸ **RV.Normal(mean,stddev)** appears. Click on this option to move **RV.Normal (?,?)** into the "Numeric Expresion:" box.
4. Click on the first "**?**" and change it to ***50*** (for the mean), and then click on the second "**?**" and change it to 15 (for the standard deviation). Type a variable name in the "Target Variable:" box. The variable name ***Out1*** is used in this example (see Figure 9.5).

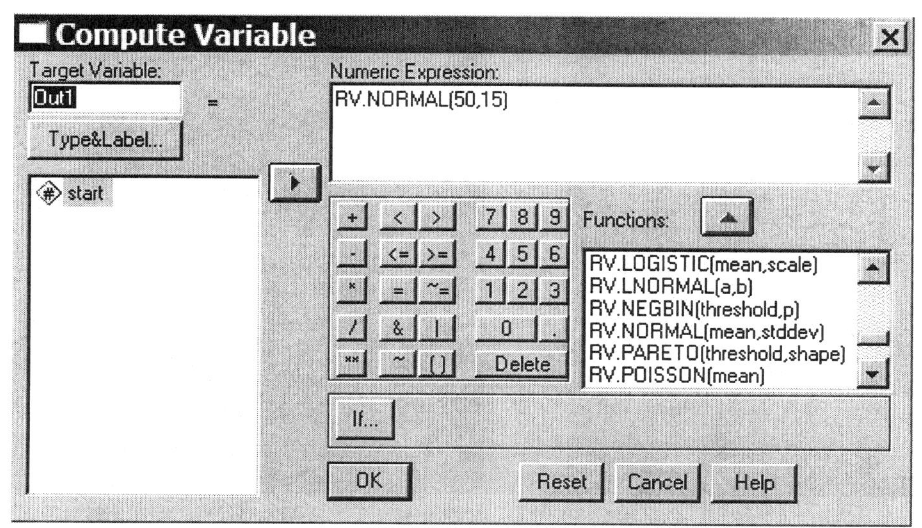

Figure 9.5

5. Click **OK.**
6. Repeat this process until you have 10 samples. In the examples that follow, the variables are called ***Out2*** to ***Out10.***
7. Calculate **Descriptives** for ***Out1*** to ***Out10*** (see Table 9.2).

Descriptive Statistics

	N	Minimum	Maximum	Mean	Std. Deviation
OUT1	50	23.66	89.34	51.4369	14.27101
OUT2	50	19.34	83.84	48.3043	13.87696
OUT3	50	25.25	83.20	48.9557	13.02511
OUT4	50	20.03	89.63	51.2718	14.76240
OUT5	50	23.43	71.39	49.8858	10.88076
OUT6	50	12.35	83.65	49.9775	16.61285
OUT7	50	12.13	83.75	51.0161	15.99582
OUT8	50	14.05	84.23	49.8595	14.97680
OUT9	50	18.16	91.61	54.6250	14.22789
OUT10	50	23.54	77.80	50.7405	13.49860
Valid N (listwise)	50				

Table 9.2

Notice that the minimum and maximum scores are quite different from sample to sample; however, the mean centers around 50 and the standard deviation centers around 15. The Dean and Registrar of your faculty would be pleased with these 10 sets of grades. Note that you can see the individual grades of each "class member" by scanning the SPSS for Windows Data Editor content that you generated. Also, because these are random samples, your outcome will be different from that shown here and from that of your classmates. However, overall, you will all have a mean of approximately 50 and a standard deviation of approximately 15 for each sample that is randomly generated.

You can use several other types of distributions to generate random samples with specific characteristics. We have generated two — one for a dichotomous variable and one for a variable that is normally distributed. If you scan the "**RV.**" options in the **Transform** ‣ **Compute** ‣ **Functions:** box, you will notice that there are a total of 22 options. Using the **Help** command, you can determine what information is needed in the prompt boxes **(?)** boxes. Explore these if you care to.

Shown below are some examples, taken from the SPSS help file for random variables (RV). Pressing the Help key while in the "Compute Variables:" window will take you to these and other options.

The following functions give a random variate from a specified distribution. The arguments are the parameters of the distribution. You can repeat the sequence of pseudo-random numbers by setting a seed in the Preferences dialog box before each sequence. Note the period in each function name.

NORMAL(stddev) Numeric. Returns a normally distributed pseudo-random number from a distribution with mean 0 and standard deviation stddev, which must be a positive number. You can repeat the sequence of pseudo-random numbers by setting a seed in the Preferences dialog box before each sequence.

RV.BERNOULLI(prob) Numeric. Returns a random value from a Bernoulli distribution with the specified probability parameter prob.

RV.BINOM(n, prob) Numeric. Returns a random value from a binomial distribution with the specified number of trials and probability parameter.

RV.CHISQ(df) Numeric. Returns a random value from a chi-square distribution with the specified degrees of freedom df.

RV.EXP(shape) Numeric. Returns a random value from an exponential distribution with the specified shape parameter.

RV.LNORMAL(a, b) Numeric. Returns a random value from a log-normal distribution with the specified parameters..

RV.NORMAL(mean, stddev) Numeric. Returns a random value from a normal distribution with the specified mean and standard deviation.

RV.UNIFORM(min, max) Numeric. Returns a random value from a uniform distribution with the specified minimum and maximum. Also see the UNIFORM function.

Chapter 10. Sampling Distributions

Topics covered in this chapter:

- **The Central Limit Theorem**
- **Control Charts**

A probability distribution for a sample statistic is often called a **sampling distribution.** This chapter is closely tied with earlier chapters that describe how to simulate random samples from a known population and compute sample statistics for the generated samples. The generated samples — for example, those we generated in Chapter 9 — can be used to examine properties of sampling distributions.

The Central Limit Theorem

Example 10.7 in BPS looks at the central limit theorem using information about how long it takes a technician to service an air conditioning unit. The time for this task is exponentially distributed with a mean $\mu = 1$ and a standard deviation $\sigma = 1$. The distribution is not normally distributed but rather has a right skew. Figure 10.4 in BPS illustrates the central limit theorem for this distribution. The following steps outline how to use SPSS to illustrate the central limit theorem.

First, we need to generate a large number of large samples. The definitions of large in this context will be set arbitrarily as 25 random samples of size n = 250. We learned how to generate such samples in Chapter 9. We will repeat the process here using an exponential distribution.

To generate 25 exponentially distributed random samples for samples of size 250 with a mean of 1 and a standard deviation of 1, follow these steps:

1. First, we need a "starting" number. In the first column of your data set, type in a number of your choosing and repeat the number until it occurs 250 times in that column. The number 37 was chosen for this example and the variable was named ***start.***
1. Click **Transform** ‣ **Compute** ‣ and the "Compute Variable" window appears.
2. Scroll down the "Functions:" box until ‣ **RV.EXP(scale)** appears. Click on this option to move **RV.EXP(?)** into the "Numeric Expresion:" box.
2. Click on the "**?**" and change it to ***1.*** Type a variable name in the "Target Variable:" box. In the example, the variable name ***Out1*** was used.
3. Click **OK.**
4. Repeat this process until you have 25 samples called ***Out1*** to ***Out25.***

Calculate **Descriptives** for ***Out1*** to ***Out25.*** A sample output is shown in Table 10.1. Recall that because these are random samples, your output will differ each time you generate the sample. Notice from Table 10.1 that the mean and standard deviation for our samples is approximately 1, which is what we asked for.

Descriptive Statistics

	N	Minimum	Maximum	Mean	Std. Deviation
OUT1	250	.00	9.23	1.0354	1.03968
OUT2	250	.01	5.43	1.0118	.93475
OUT3	250	.00	6.23	.9852	.98888
OUT4	250	.00	7.54	1.1109	1.13951
OUT5	250	.00	5.92	1.0037	1.03621
OUT6	250	.00	7.65	.9880	1.02262
OUT7	250	.00	5.30	.9612	.98417
OUT8	250	.01	5.88	.9629	.89949
OUT9	250	.01	4.52	1.0318	.91248
OUT10	250	.00	4.22	.9588	.86639
OUT11	250	.01	5.56	.9614	1.01616
OUT12	250	.01	4.49	.9702	.93340
OUT13	250	.00	4.34	.9608	.98151
OUT14	250	.00	6.93	1.0345	1.00478
OUT15	250	.01	5.99	1.0954	1.02468
OUT16	250	.00	5.16	.9966	.92703
OUT17	250	.01	4.81	1.0546	.96721
OUT18	250	.00	5.94	.9689	.96395
OUT19	250	.01	8.54	1.1464	1.26292
OUT20	250	.00	9.09	1.0274	1.04345
OUT21	250	.01	6.34	.8974	.93751
OUT22	250	.00	7.87	1.0218	1.08720
OUT23	250	.00	5.25	.9915	.96127
OUT24	250	.00	4.86	.9813	1.00465
OUT25	250	.02	8.42	1.0374	1.09938
Valid N (listwise)	250				

Table 10.1

In Figure 10.1, we can see the distribution of repair times for our 25th sample, ***out25.*** When doing the histogram, if you click in the "Display Normal Curve" box, SPSS will show the normal curve over the histogram for the sample. In this case, there is a strong right skew as we expected from the characteristics of exponential curves. Each of our 25 samples will take approximately this shape.

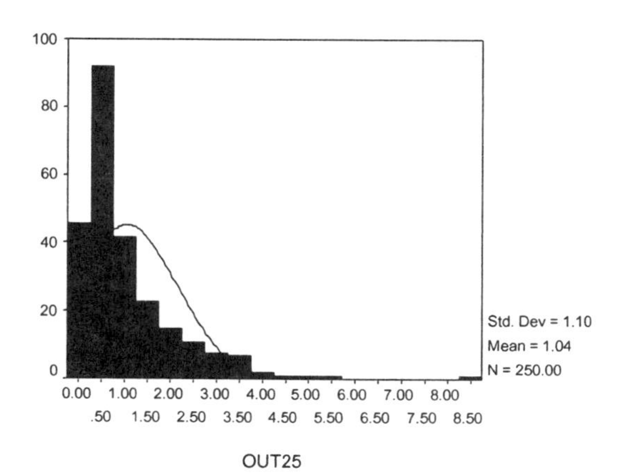

Figure 10.1

Each individual mean shown in Table 10.1 is a sample statistic. To generate a *sampling distribution*, we need to create an SPSS spreadsheet showing each mean and standard deviation for this set of samples. Take time now to create new numeric variables called ***mean*** and ***stddev*** or other variable labels that are meaningful to you. Enter the mean for each of your 25 samples under the ***mean*** and the standard deviation for each of your 25 samples under ***stddev.*** You can use the copy and paste functions to do this. An example for the first 10 means and standard deviations is shown in Figure 10.2.

Descriptive Statistics

	N	Minimum	Maximum	Mean	Std. Deviation
OUT1	250	.00	9.23	1.0354	1.03968
OUT2	250	.01	5.43	1.0118	.93475
OUT3	250	.00	6.23	.9852	.98888
OUT4	250	.00	7.54	1.1109	1.13951
OUT5	250	.00	5.92	1.0037	1.03621
OUT6	250	.00	7.65	.9880	1.02262
OUT7	250	.00	5.30	.9612	.98417
OUT8	250	.01	5.88	.9629	.89949
OUT9	250	.01	4.52	1.0318	.91248
OUT10	250	.00	4.22	.9588	.86639

Untitled - SPSS Data Editor

File Edit View Data Transform Analyze Graphs Utilities Window Help

1 : stdev 1.03968

	out24	out25	mean	stdev	var	var
1	.13	1.16	1.04	1.04		
2	.91	.16	1.01	.93		
3	.26	.47	.99	.99		
4	3.80	.81	1.11	1.14		
5	.42	.75	1.00	1.04		
6	2.85	1.39	.99	1.02		
7	.69	.93	.96	.98		
8	.19	.58	.96	.90		
9	.08	.04	1.03	.91		
10	1.98	.13	.96	.87		

Figure 10.2

Next, do descriptive statistics for these *sampling* distributions. See a sample output in Table 10.2.

Descriptive Statistics

	N	Minimum	Maximum	Mean	Std. Deviation
MEAN	25	.90	1.15	1.0078	.05451
STDEV	25	.87	1.26	1.0016	.08425
Valid N (listwise)	25				

Table 10.2

Notice that we now have the mean of the means = 1.0078 (still labeled by SPSS as **Mean**) and the standard error of the sampling distribution of the means = .05451 (still labeled by SPSS as **Std. Deviation**). We also have the mean of the standard deviations = 1.0016 and the standard deviation of the sampling distribution of the standard deviations = .08425.

When we draw a histogram of the distribution of the means (sampling distribution of the mean) as shown in Figure 10.3, it approximates the shape of a normal curve. Graph the sampling distribution of the standard deviations with a normal curve superimposed and you will get a similar result (see Figure 10.4).

The larger the number of samples, the closer the sampling distributions will come to normal curves. If you have the time, continue the process of generating random samples until you have 100 rather than 25 and then look at the sampling distribution of both the mean and the standard deviation.

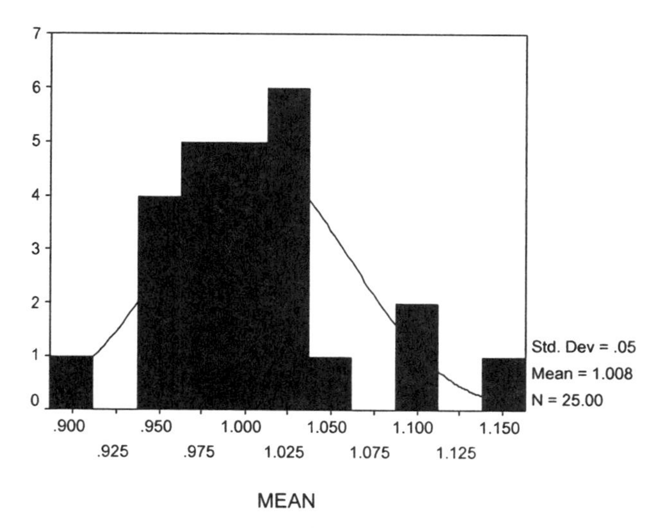

Figure 10.3

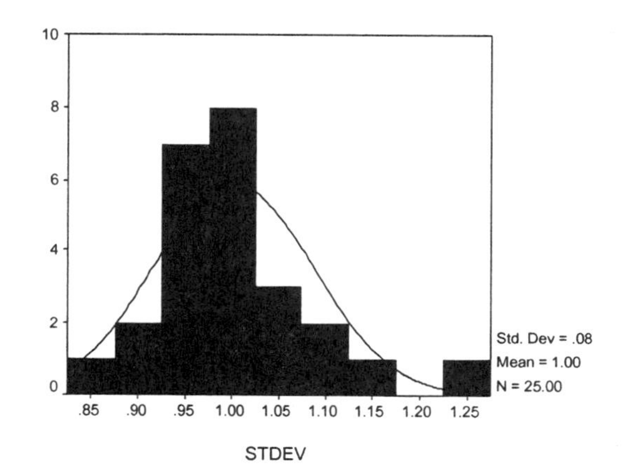

Figure 10.4

Control Charts

Control Charts allow you to monitor processes, such as production of a manufactured article and how the production meets or fails to meet production specifications over time. Anomalies in the control charts may be good indicators of some problem in the production process. We often use the standard deviation or multiples of it **(sigmas)** to define acceptable ranges for these processes. When the process is "out of range," it may prompt us to review the situation to make changes to bring the outcome back within range.

Example 10.8 of BPS presents data from a manufacturer of computer monitors. The manufacturer measures the tension of fine wires behind the viewing screen. These measurements are taken four times every hour. The ideal tension for these wires is 275 mV. When the process is operating properly, the standard deviation of the tension readings is $\sigma = 43$ mV.

Table 10.1 of BPS and **ta10-01.por** contain the measurements for 20 hours. In this data set, each row contains the sample number, the four sample values, and the sample mean. To obtain the control chart for each individual measurement in the mean (**v6**) column, follow these steps:

1. Click **Graphs** ‣ **Control** ‣ **Individuals** ‣ **Define.** The "Individuals and Moving Range" window appears as in Figure 10.5.
2. Click ***v6,*** then click ‣ to move ***v6*** into the "Process Measurement:" window.
3. Click on **individuals.** If you wish to input your own upper and lower control limits, click **statistics** and enter them in the appropriate boxes. Otherwise, SPSS will use + and - three sigmas as the default upper and lower control limits. Under **options** you can change the number of **sigmas** from three to either one or two.
4. Click OK.

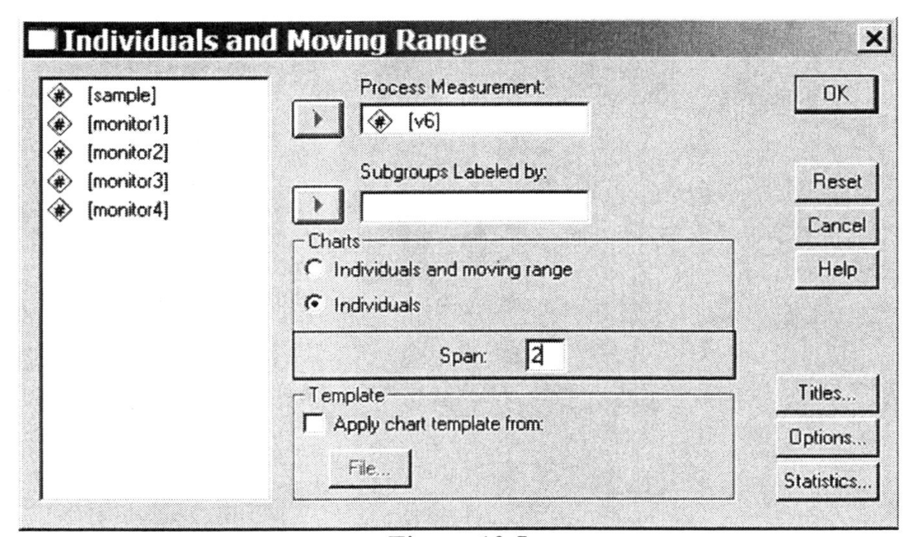

Figure 10.5

The control chart produced and shown in Figure 10.6, includes a moving line with both upper and lower control limits. Each point on the process line represents an individual case. The upper control limit in this example is set at the mean plus 3 standard deviations (**sigmas**) and the lower control limit is set at minus three standard deviations. These values, along with the overall mean are shown at the lower right of the chart. The sigma level is shown at the lower left.

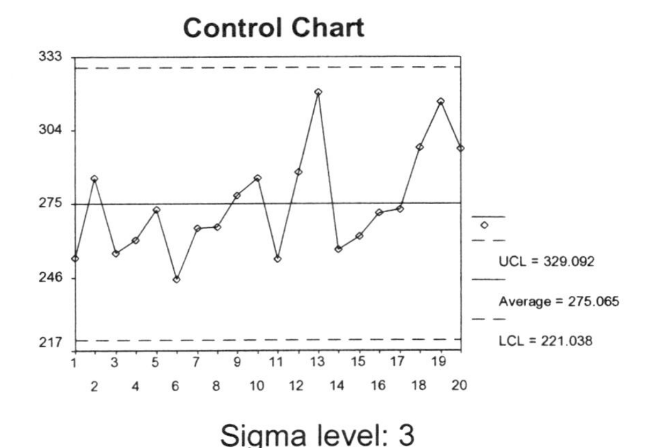

Figure 10.6

Chapter 11. General Rules of Probability

Topic covered in this chapter:

- **Probability Calculations**

Probability Calculations

There is a calculator within SPSS that can be used to do arithmetic calculation such as those needed for the basic rules of probability. The calculator performs basic arithmetic operations, such as addition (+), subtraction (-), multiplication (*), division (/), and exponentiation (**). These operations can be used for calculations based on the probability rules described in BPS. You can type the function in or choose it from the function box. The "Compute Variable" window can be accessed by clicking **Transform** then **Compute.**

In order to use the calculator, it is necessary to have some data in the SPSS spreadsheet. To illustrate the process, consider Example 11.2 in BPS. This example describes British bomber missions where the probability of losing the bomber in a mission was .05. The probability that the bomber returned therefore was 1 - .05 = .95. It seems reasonable to assume that missions are independent. Therefore, the probability of surviving 20 missions is:

$$P(A_1 \text{ and } A_2 \text{ and } \ldots \text{ and } A_{20}) = P(A_1)*P(A_2)* \ldots *P(A_{20}) = (.95)^{20}$$

To calculate this probability, first enter the number .95 in the first cell of an SPSS spreadsheet and then complete the following steps:

1. Click **Transform** then **Compute.**
2. In the "Compute Variable" window, type in the name of the "Target Variable:"
3. Click on the existing variable, then click ‣ so that the existing variable moves into the "Numeric Expression:" window.
4. Following the variable name in the "Numeric Expression:" window, type in *****20*** (meaning raised to the power of 20 or multiplied by itself 20 times) (see Figure 11.1).
5. Click **OK.**

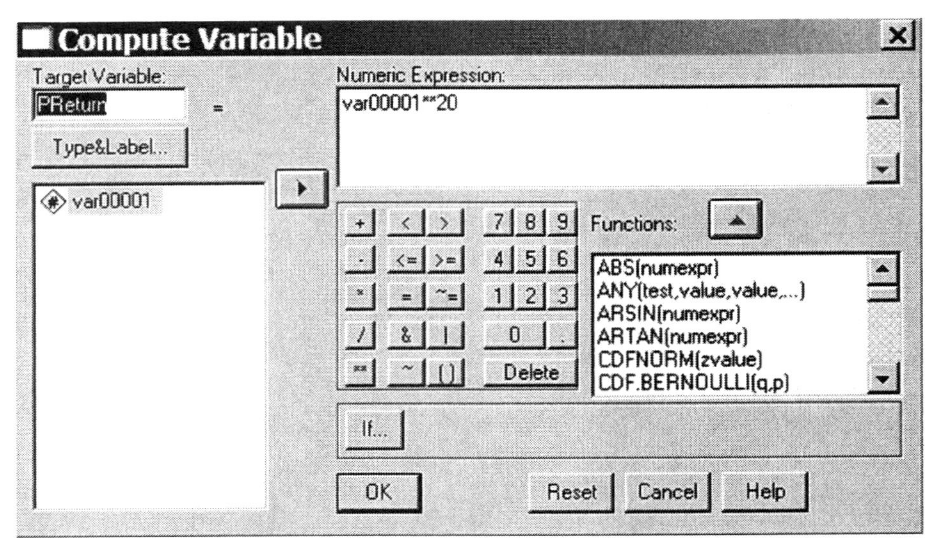

Figure 11.1

The result, Shown in Figure 11.2, is that the probability of returning from 20 missions is approximately .36.

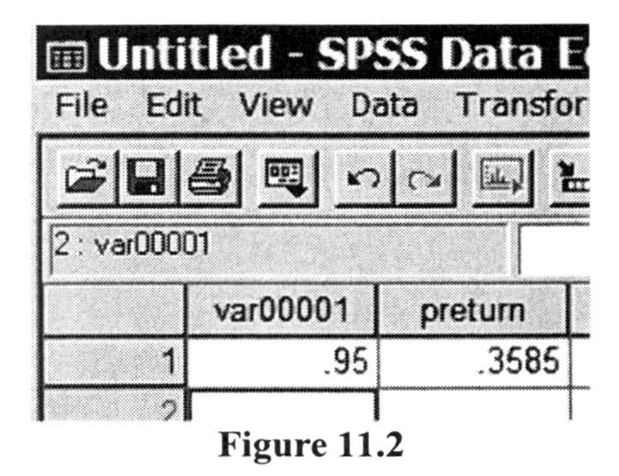

Figure 11.2

This calculator can be used for any of the expressions used to compute conditional probabilities and addition and multiplication rules. Using the examples in BPS Chapter 11, use the calculator to replicate the outcomes there.

Chapter 12. Binomial Distributions

Topics covered in this chapter:

- **Binomial Probabilities**
- **Normal Approximations to the Binomial**

This chapter will show how SPSS for Windows can be used to find **probabilities** associated with a **binomial probability distribution.** In addition, a list of other probability distributions for which SPSS for Windows can calculate probabilities is given at the end of the chapter.

In BPS Example 12.6, an engineer chooses a simple random sample of 10 switches from a shipment of 10,000 switches. Suppose that, unknown to the engineer, 10% of the switches in the shipment are bad. The engineer counts the number X of bad switches in the sample. What is the probability that no more than 1 of the 10 switches in the sample fails inspection?

Let X = the number of switches in the sample that fail to meet the specifications. X is a binomial random variable with $n = 10$ and $p = 0.10$. To find the probability that no more than 1 of the 10 switches in the sample fails inspection ($P(X \leq 1)$ when $n = 10$ and $p = 0.10$), follow these steps:

1. Define a new variable (e.g., ***switch***), which takes on the values 0 through 10. This variable was declared a numeric variable of length 8.0.
2. Click **Transform** and then click **Compute.** The "Compute Variable" window in Figure 12.1 appears.

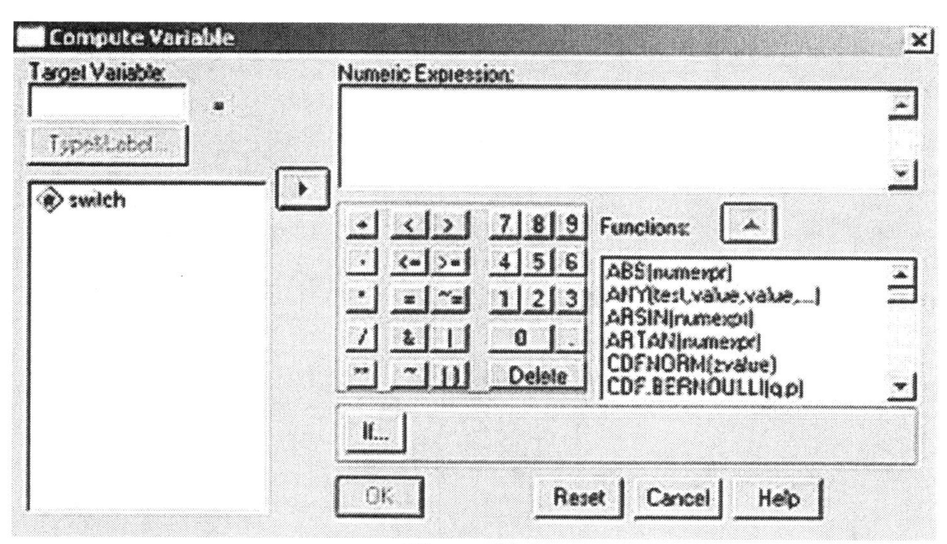

Figure 12.1

3. In the "Target Variable" box, type in ***lesseq.***
4. In the "Functions" box, click ▾ until **CDF.BINOM(q,n,p)** appears in the box. Double-click on **CDF.BINOM(q,n,p)** to move **CDF.BINOM(?,?,?)** into the "Numeric Expression" box. The **CDF.BINOM(q,n,p)** function stands for the cumulative distribution function for the binomial

distribution, and it calculates the cumulative probability that the variable takes a value less than or equal to q.

5. In the "Numeric Expression" box, highlight the first ***?***, click on ***switch,*** and click ▸ so ***switch*** replaces the first **?**. Replace the second **?** with the number **10** (the value of n), then replace the third **?** with the value **0.10** (the value of p).
6. Click **OK.**

Figure 12.2 displays the new variable ***lesseq.*** By default, SPSS shows the values of ***lesseq*** to two decimal places of accuracy. The number of decimal places was changed to four.

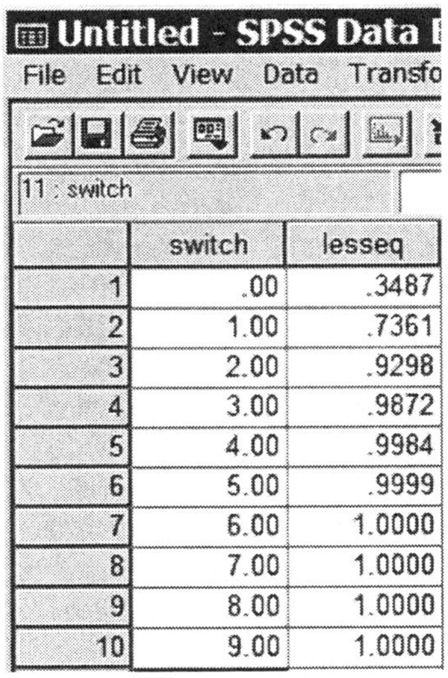

Untitled - SPSS Data E

File Edit View Data Transfo

11 : switch

	switch	lesseq
1	.00	.3487
2	1.00	.7361
3	2.00	.9298
4	3.00	.9872
5	4.00	.9984
6	5.00	.9999
7	6.00	1.0000
8	7.00	1.0000
9	8.00	1.0000
10	9.00	1.0000

Figure 12.2

For each value of ***switch,*** the variable ***lesseq*** represents the cumulative probability of observing that number or fewer failures.

We want to determine the probability that no more the 1 of the 10 switches in the sample fails inspection or, symbolically, $P(X \leq 1 \mid n = 10, p = 0.10)$. The variable ***lesseq*** tells us that the probability that no more than 1 of the 10 switches in the sample fails inspection is 0.7361, given that the probability any given switch will fail is $p = 0.10$.

Given the same values of n and p, suppose one wants to find the probability that strictly less than two switches fail inspection, $P(X < 2 \mid n = 10, p = 0.10)$. Steps 2 to 6 can be repeated, with the changes that the target variable is named ***less*** and the "Numeric Expression" box reads **CDF.BINOM(switch-1,10,0.10).** If one wants to determine the probability that exactly two of the 10 switches fail inspection, $P(X = 2 \mid n = 10, p = 0.10)$, steps 2 to 6 can be repeated, with the changes that the target variable is named ***equal*** and the "Numeric Expression" box reads **CDF.BINOM(switch,10,0.10) - CDF.BINOM(switch-1,10,0.10).**

To determine the probability that at least two of the 10 switches fail inspection, $P(X \geq 2 \mid n = 10, p = 0.10)$, steps 2 to 6 can be repeated, with the changes that the target variable is named ***greateq*** and the "Numeric Expression" box reads **1 - CDF.BINOM(switch-1,10,0.10).**

Finally, if one wants to determine the probability that strictly more than two of the 10 switches fail inspection, $P(X > 2 \mid n = 10, p = 0.10)$, steps 2 to 6 can be repeated, with the changes that the target variable is named ***greater*** and the "Numeric Expression" box reads **1 - CDF(switch,10,0.10).** The probabilities associated with the first three switches are shown below in Figure 12.3. More specifically, Row 3 shows the probabilities associated with $X \leq 2$, $X < 2$, $X = 2$, $X \geq 2$, and $X > 2$ under the variables of ***lesseq, less, equal, greateq,*** and ***greater,*** respectively.

	switch	lesseq	less	equal	greateq	greater
1	.00	.35	.00	.35	1.00	.65
2	1.00	.74	.35	.39	.65	.26
3	2.00	.93	.74	.19	.26	.07

Figure 12.3

NOTE: The formats needed to obtain these various binomial probabilities are summarized in Table 12.1. The symbol x is used to represent some value of X. For instance, $P(X < x)$ could represent the expression $P(X < 5)$ or the expression $P(X < 12)$.

Binomial Probability Sought	CDF. BINOM Format Needed
$P(X < x)$	CDF.BINOM(x – 1,n,p)
$P(X \leq x)$	CDF.BINOM(x,n,p)
$P(X = x)$	CDF.BINOM(x,n,p) – CDF.BINOM(x – 1,n,p)
$P(X \geq x)$	1 – CDF.BINOM(x – 1,n,p)
$P(X > x)$	1 – CDF.BINOM(x,n,p)

Table 12.1 Binomial Probability Calculations

An opinion poll described in BPS Example 12.7 asks 2500 adults whether they agree or disagree that "I like buying new clothes, but shopping is often frustrating and time consuming." Suppose that 60% of all U.S. residents would say "Agree."

To find $P(X \geq 1520)$, the probability that at least 1520 adults agree, follow the steps as outlined in Table 12.1. Create a variable named ***agree*** and enter the single value ***1520***. Use the steps outline above until the "Compute Variable" Window appears. Give your new variable a name and insert the following in the "Numeric Expression" box: **1-(CDF.BINOM[agree,2500,.6])**.

SPSS returns the value of .20, meaning that the probability that 1520 people will agree is .20.

SPSS for Windows is capable of computing probabilities for a number of distributions. Table 12.2, shown on the next page, displays a number of commonly used distributions and their commands.

Table 12.2

Distribution	SPSS Command
Chi-square	CDF.CHISQ(q,df)
Exponential	CDF.EXP(q,scale)
F	CDF.F(q,df1,df2)
Geometric	CDF.GEOM(q,p)
Hypergeometric	CDF.HYPER(q,total,sample,hits)
Normal	CDF.NORMAL(q,mean,stddev)
Poisson	CDF.POISSON(q,mean)
Uniform	CDF.UNIFORM(q,min,max)

Chapter 13. Confidence Intervals: The Basics

Topics covered in this chapter:

- **Simulating Repeated Random Samples**
- **One-Sample Z Confidence Interval**

A **confidence interval** is a procedure for estimating a population parameter using observed data. This chapter describes how to **simulate** random samples from a known population and examine various **properties of confidence intervals.**

Simulating Repeated Random Samples

A simulation can be used to demonstrate how a confidence interval works:

1. Open your internet browser and go to: http://bcs.whfreeman.com/bps3e/
2. Under **Student Resources,** click on **Statistical Applets.** Then click on **Confidence Interval.**
3. Follow the instructions on the screen to generate a picture like the one in BPS Figure 13.4.

What happens as you add one sample at a time? Fifty samples at a time? What does this visual representation tell you about sampling *in the long run*? What Theorem does this representation illustrate?

One-Sample Z Confidence Interval

Exercise 13.6 in BPS gives you the IQ scores for 31 seventh-grade girls in a Midwest school district. Open these data from your **PCDataSets.** They are called **ex13-06.por.** We are asked to verify that there are no major departures from normality and to obtain the 99% confidence interval for the IQ of the population of seventh-grade girls. We are given $\sigma = 15$.

To calculate the confidence interval using SPSS, follow these steps:

1. Click **Analyze** ‣ **Descriptive Statistics** ‣ **Explore.**
2. Click on ***iq*** and then ‣ to move ***iq*** into the "Dependent List" window.
3. Click "Plots" in the lower right corner, then click on "Normality Plots with Tests."
4. Click **Continue.**
5. Click **OK.**

To verify that there are no major departures from normality, check the stem-and-leaf for outliers and then look at the Normal Q-Q Plot. The Normal Q-Q plot for these ***iq*** data is shown in Figure 13.1. If all the points are near the straight line, as they are in this example, we have good evidence for a normal distribution in this type of visual presentation.

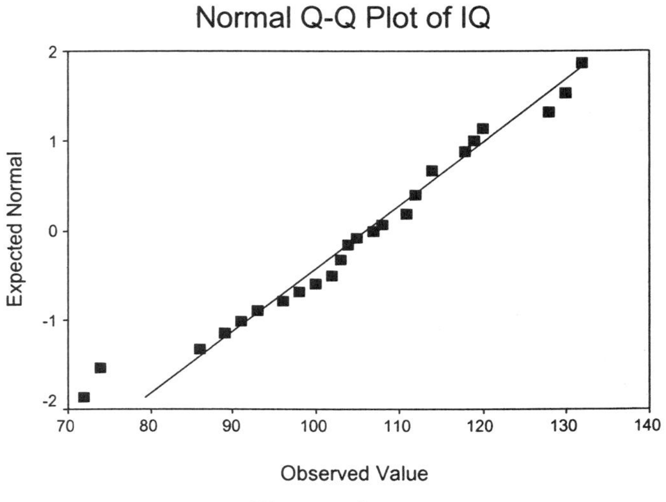

Figure 13.1

Information required to compute the 95% confidence interval can be found in the Descriptives output shown in Table 13.1. The confidence interval noted in the Descriptives table is for $\bar{x}$.

Descriptives

		Statistic	Std. Error
Mean		105.84	2.563
95% Confidence Interval for Mean	Lower Bound	100.60	
	Upper Bound	111.07	
5% Trimmed Mean		106.27	
Median		107.00	
Variance		203.673	
Std. Deviation		14.271	
Minimum		72	
Maximum		132	
Range		60	
Interquartile Range		16.00	
Skewness		-.470	.421
Kurtosis		.431	.821

Table 13.1

To compute the confidence interval of μ, follow the steps outlined below and reference BPS Chapter 13:

1. Compute the standard error of the $\bar{x}$ by dividing σ by the sqrt of n. In SPSS, enter the n and σ into a spreadsheet and use the **Transform** ‣ **Compute** features. See Figure 13.2 for an example.
2. Click OK and the value 2.69 is returned.

For a 95% confidence interval, we need $\bar{x} \pm 2$ standard deviations.

Thus, we have 105.84 – 2*2.69 = 100.46 and 105.84 + 2*2.69 = 111.22.

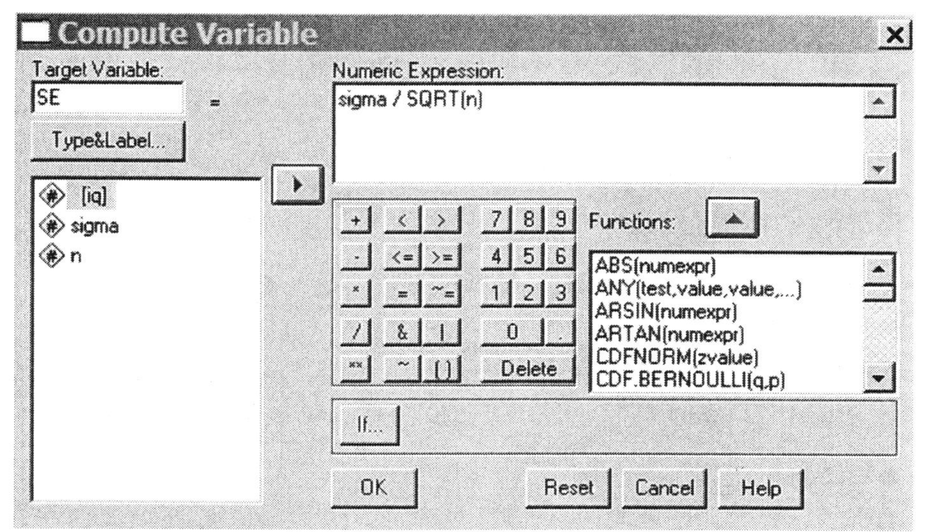
Compute Variable
Target Variable:
SE
=
Type&Label...
Numeric Expression:
sigma / SQRT(n)
[iq]
sigma
n
Functions:
ABS(numexpr)
ANY(test,value,value,...)
ARSIN(numexpr)
ARTAN(numexpr)
CDFNORM(zvalue)
CDF.BERNOULLI(q,p)
Delete
If...
OK
Reset
Cancel
Help

Figure 13.2

Chapter 14. Tests of Significance: The Basics

Topic covered in this chapter:

- **One-Sample Z Test**

A **test of significance** is a procedure for determining the validity of a claim using observed data. This chapter describes how to examine various properties of tests of significance.

One-Sample Z Test

Example 14.2 in BPS describes a test for sweetness loss in artificially flavored colas. Trained tasters sip the cola then sip drinks of standard sweetness and then score the cola on a "sweetness score" from 1 to 10. Artificial sweeteners gradually lose their sweetness over time so the colas are stored at a high temperature for one month and then are tested and again rated for sweetness. This is a matched pairs experiment and the changes in the sweetness score (score before minus score after) form the data. The bigger the difference from before to after, the greater the sweetness loss. Are these data good evidence that the colas lost sweetness during storage? In order to answer this question, we must test the hypothesis that $\mu = 0$ (no sweetness loss) against the hypothesis that $\mu = 0$. You can find these difference scores in the SPSS PCDataSets for Chapter 14 by opening **eg14-02.por.**

To test your hypotheses, first obtain the mean and standard deviation for your data and check for outliers and normality as described in Chapter 13 of this manual. The steps are repeated for you below.

1. Click **Analyze** ▸ **Descriptive Statistics** ▸ **Explore.**
2. Click on **Sweetness** and then ▸ to move **Sweetness** into the "Dependent List" window.
3. Click "**Plots**" in the lower right corner then click on "Normality Plots with Tests."
4. Click **Continue.**
5. Click **OK.**

To verify that there are no major departures from normality, check the stem-and-leaf for outliers and then look at the Normal Q-Q Plot to see if all of the points are near the line. The SPSS output shows no reason for concern about outliers or normality. Confirm this in your own analysis of these data.

Next we need to compute the standard error again as we did in Chapter 13. See the steps below:

1. Compute the standard error of the $\bar{x}$ by dividing σ by the sqrt of n. In SPSS, enter the n (**10**) and σ (***1***) into a SPSS for Windows Data Editor spreadsheet and use the **Transform** ▸ **Compute** features (see Figure 14.1).
2. Click **OK** and the value .3162 is returned which matches the outcome in BPS.

To complete our test of the hypotheses, it is necessary now to calculate a z statistic. The z statistic tells us how far $\bar{x}$ is from μ in standard deviation units. To calculate the z statistic and its associated p value, follow the instructions on the next page.

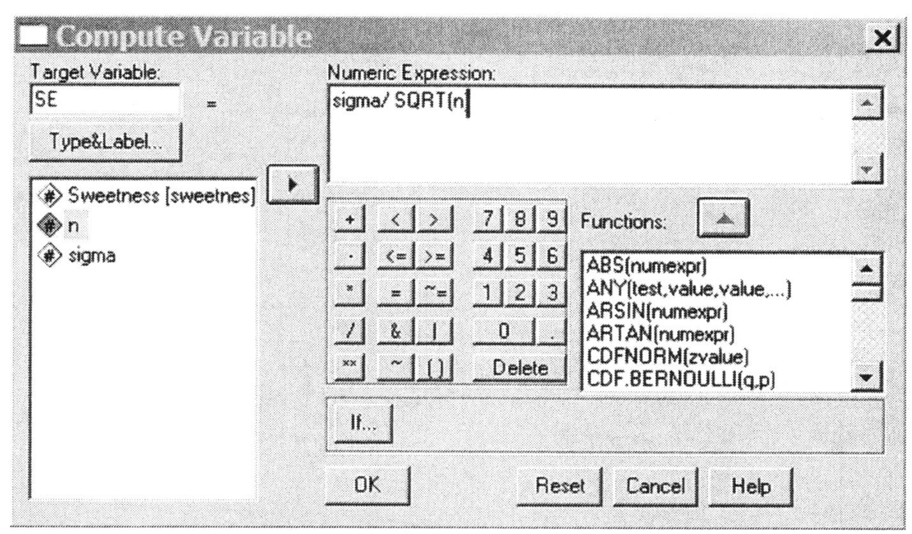

Figure 14.1

To calculate the z statistic:

1. Enter the sample mean, $\bar{x}$, in the SPSS for Windows Data Editor.
2. Using the **Transform** ‣ **Compute** features, calculate z = ($\bar{x}$ - μ)/SE (see Figure 14.2).
3. Click **OK** and SPSS will return the value 3.23 as in BPS Example 14.4.

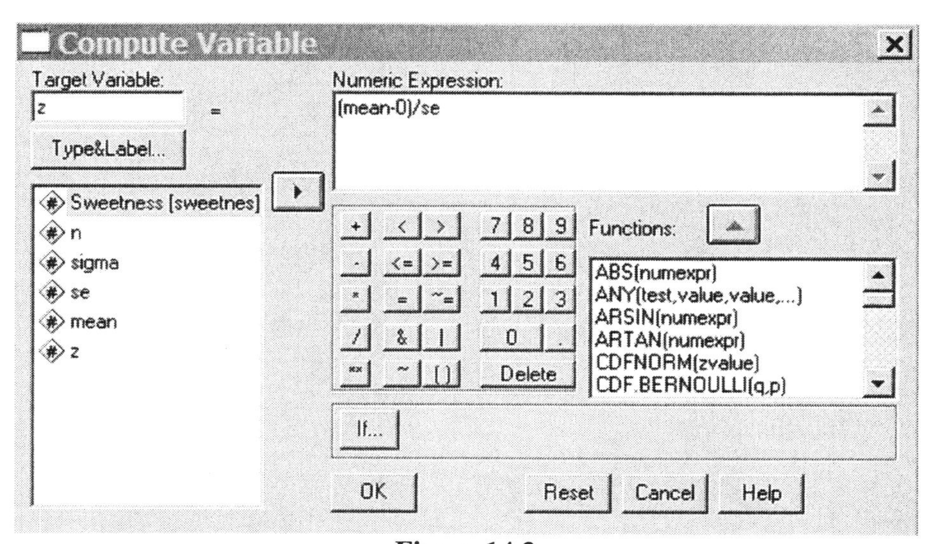

Figure 14.2

To calculate the probability:

To finish the process of testing our hypotheses, we now need to calculate the probability of getting a z at least this far from zero in either direction. To do this we need P = P(Z > 3.23). You may wish to review the probability calculations in Chapters 2 and 11 of this manual. The calculations for this problem are illustrated below and in Example 14.5 of BPS.

To calculate the probability of getting a z value less than 3.23, use the **CDF.NORMAL** function in SPSS. To get this value, use the following steps:

1. Click on **Transform** ‣ **Compute** and the "Compute Variable" window will appear.
2. Label your target variable (***p***, for example).
3. Choose the **CDF.NORMAL** function by locating it in the "Functions:" window and then double-clicking it to move it into the "Numeric Expression:" window.
4. Replace the first **?** with the ***value for z*** or the ***variable name*** for the location of z in your SPSS for Windows Data Editor. Replace the second **?** with the mean of a z distribution (***0***) and replace the third **?** with the value of the standard deviation for the z distribution (***1***) (see Figure 14.3).
5. Click **OK.** SPSS will return the value .99937, which can be rounded to .9994 to match the value shown in Example 14.5 in BPS.

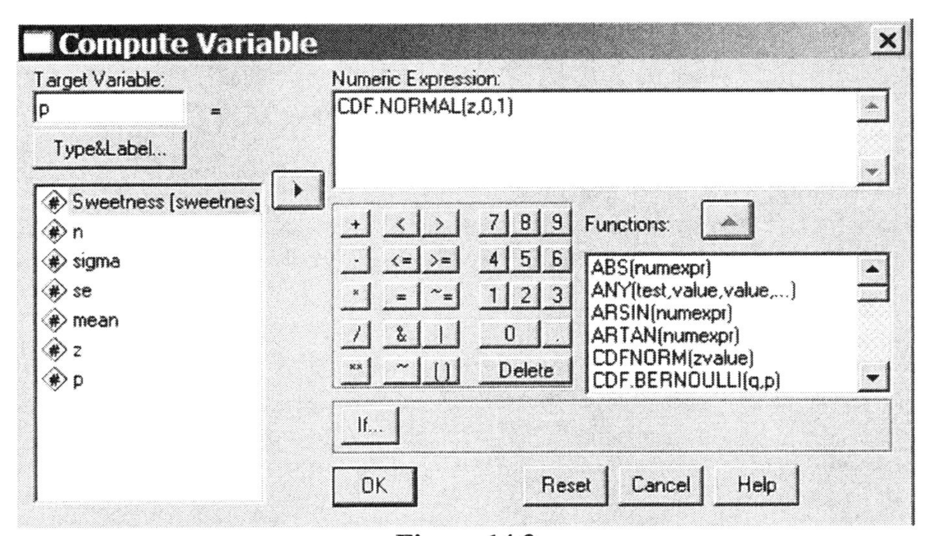

Figure 14.3

Recall that CDF means a Cumulative Distribution Functions and therefore gives the probability of every z value less than 3.23 for this example. Look at Figure 14.2 in BPS to remind yourself that what you want is the probability of z equal to or greater than 3.23. Therefore, to complete our calculations we must do the following:

$$P = P(Z > 3.23) = 1 - .9994 = .0006.$$

This small probability gives us strong evidence *against* H_0 and in favor of H_a.

Chapter 15. Inference in Practice

Topic covered in this chapter:

- **Calculating Power**

Power represents your ability to reject a false null hypothesis in favor of the alternative. Power is of interest to scientists analyzing the data that result from their experiments because generally they are interested in rejecting the null hypothesis. To illustrate the concept of power, we will return to the cola sweetness example from Chapter 14.

Calculating Power

Example 15.5 in BPS illustrates the power calculations for the cola sweetness example. SPSS can be used to examine the power of the test against the specific alternative of $\mu = 0$. The test rejects H_o when the test statistic z is ≥ 1.645 or when the sample mean $\bar{x}$ is $\geq .520$.

The calculations for power are done in two steps.

For Step 1, we are asked to write the rule for rejecting H_0 in terms of the $\bar{x}$. We know that $\sigma = 1$, so the z test rejects H_0 at the $\alpha = .05$ level when $\bar{x}$ is $\geq .520$. We can use SPSS to generate this value of .520 as follows:

1. Click **Transform, Compute** and in the "Compute Variable" window, label your "Target Variable" (e.g. ***sampmean***), then enter the calculation equation as shown in Example 15.5 in the "Numeric Expression:" window (see Figure 15.1).
2. Click **OK.**

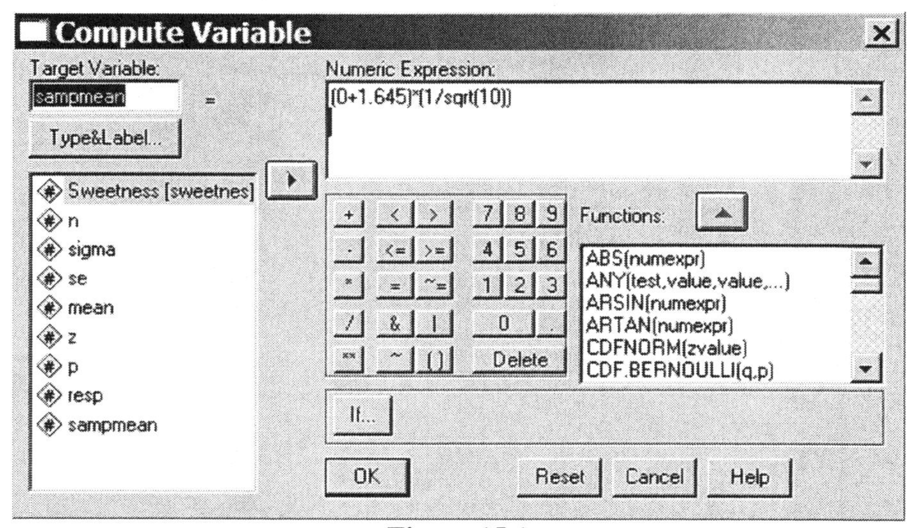

Figure 15.1

In Step 2, we calculate power. Power is the probability of this event under the condition that the alternative $\mu = 1.1$ is true. To calculate this we need to standardize $\bar{x}$ using $\mu = 1.1$. To calculate power, follow these steps:

1. Click **Transform, Compute** and in the Compute Variable window, label your Target Variable (e.g. ***prob***), then enter the calculation equation as shown in Example 15.5 in the "Numeric Expression" window (see Figure 15.2).
2. Click **OK.**

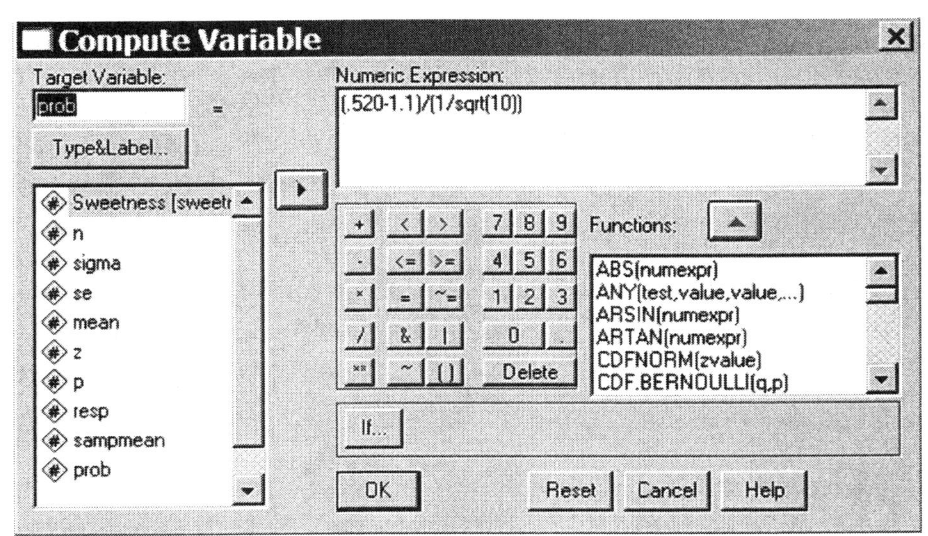

Figure 15.2

Follow the instructions in BPS Example 15.5 for using Table A to complete the problem.

Chapter 16. Inference About a Population Mean

Topics covered in this chapter:

- **One-Sample *t* Confidence Interval**
- **One-Sample *t* Procedures**
- **Matched Pairs**

This chapter introduces the use of the ***t* distribution** in inferential statistics for the mean of a population. When σ is unknown for the population, the t distribution, rather than the z distribution, is used. A particular t distribution is specified by giving the degrees of freedom. The **one-sample *t* confidence interval, one-sample *t* test for μ,** and the **matched pairs *t* procedures** are discussed in this chapter.

One-Sample *t* Confidence Interval

To study the healing of skin wounds, biologists measured the rate at which new cells closed a razor cut made in the skin of an anesthetized newt. This experiment is described in Example 16.1 in BPS, and the data can be read into SPSS from **eg16-01.por.**

Compute a 95% confidence interval for μ where μ is the mean healing rate in the population of all newts. Before proceeding with the confidence interval for the mean, we must verify the assumption of normally distributed data and compute the $\bar{x}$ and s for the data set.

The SPSS for Windows Data Editor contains a single variable called ***rate,*** which is declared type numeric 8.0.

To obtain a confidence interval for μ, follow these steps:

1. Click **Analyze,** click **Descriptive Statistics,** and then click **Explore** and the "Explore" window appears.
2. Click ***rate,*** then click ▸ to move ***rate*** into the "Dependent List" box.
3. By default, a 95% confidence interval for μ will be computed. To change the confidence level, click **Statistics**. The "Explore: Statistics" window and change 95 in the "Confidence Interval for Mean" box to the desired confidence level.
4. Click **Continue.**
5. Click "Plots" in the lower right corner and then click on "Normal Probability Plots with Tests" to obtain the Normal Q-Q plot. Review Chapter 13 of this manual for details.
6. By default, the "Display" box in the lower left corner of the "Explore" window has "Both" selected. Click **Statistics.** If you want to obtain a stemplot to check the assumption of normality, then skip this step.
7. Click **OK.**

Table 16.1 contains the resulting SPSS for Windows output. We are 95% confident that the mean healing rate lies somewhere between 21.53 and 29.81 micrometers per hour (obtained from the "Lower Bound" and "Upper Bound" rows in Table 16.1).

Descriptives

		Statistic	Std. Error
Mean		25.67	1.962
95% Confidence Interval for Mean	Lower Bound	21.53	
	Upper Bound	29.81	
5% Trimmed Mean		25.69	
Median		26.50	
Variance		69.294	
Std. Deviation		8.324	
Minimum		11	
Maximum		40	
Range		29	
Interquartile Range		12.25	
Skewness		-.253	.536
Kurtosis		-.659	1.038

Table 16.1

One-Sample *t* Procedures

To illustrate one-sample t procedures, we will return to the cola-sweetening example that continues in Example 16.2 in BPS. Cola makers test new recipes for loss of sweetness during storage. Trained tasters rate the sweetness before and after storage. Here are the sweetness losses (sweetness before storage minus sweetness after storage) found by 10 tasters for one new cola recipe: 2.0, 0.4, 0.7, 2.0, –0.4, 2.2, –1.3, 1.2, 1.1, and 2.3. Is there sufficient evidence that the cola lost sweetness at $\alpha = 0.05$?

Before proceeding with the one-sample *t* test, we must verify the assumption of normally distributed data. To obtain a Q-Q plot and/or stemplot for the variable of interest, follow the directions given above and in earlier chapters of this manual. According to a stemplot, the distribution of the 10 taste scores does not have a regular shape, but there are no gaps or outliers or other signs of nonnormal behavior.

Set up H_0: $\mu = 0$ versus H_a: $\mu > 0$, where μ = the mean sweetness loss for a large population of tasters.

These data can be retrieved from **eg16-02.por.** The SPSS for Windows Data Editor contains a single variable called ***sweetlos,*** which is declared type numeric 8.2.

To conduct a one-sample *t* test, follow these steps:

1. Click **Analyze,** click **Compare Means,** and then click **One-Sample T Test.** The "One-Sample T Test" window in Figure 16.1 appears.

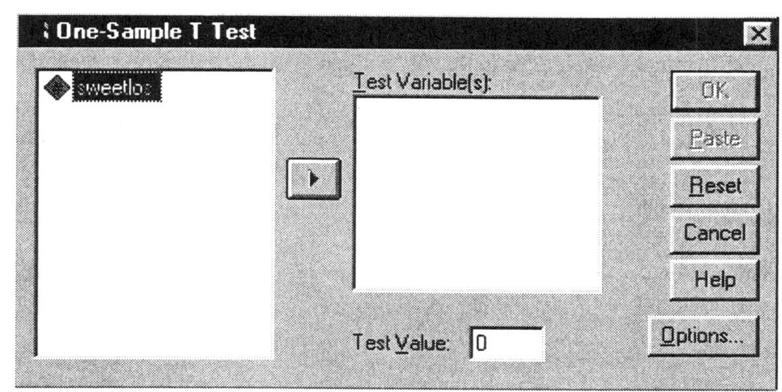

Figure 16.1

2. Click ***sweetlos,*** then click ▸ to move ***sweetlos*** to the "Test Variable(s)" box.
3. Type the value of μ_0 (the value of μ under H_0) into the "Test Value" box. For this example, 0 is the correct value. However, this will not always be the case.
4. By default, a 95% confidence interval for $\mu - \mu_0$ will be part of the one-sample t test output. To change the confidence level, click **Options,** change 95 in the "Confidence Interval" box to the desired confidence level, and then click **Continue.**
5. Click **OK.**

Table 16.2 is part of the resulting SPSS for Windows output.

One-Sample Test

	Test Value = 0					
	t	df	Sig. (2-tailed)	Mean Difference	95% Confidence Interval of the Difference	
					Lower	Upper
SWEETLOS	2.697	9	.025	1.0200	.1644	1.8756

Table 16.2

Because of the one-sided alternative, we are interested only in the upper right tail above 2.697. Therefore, the P-value is 0.0125. This is obtained by taking the value under the "Sig. (2-tailed)" column and dividing it by 2 (0.025/2). We can safely reject H_0 in favor of H_a; that is, there is sufficient evidence at the 0.05 level to conclude that the mean sweetness loss for a large population of tasters is greater than zero. In other words, there is strong evidence for a loss of sweetness.

In addition, we are 95% confident that the true mean sweetness loss lies between 0.1644 and 1.8756. *Note:* If $\mu_0 \neq 0$, then the confidence interval for μ can be obtained by adding μ_0 to the values located in the "Lower" and "Upper" boxes within the "Confidence Interval of the Difference" box (see Table 16.2).

The next example shows how to generate the exact P-value for the one-sample t test when summarized data rather than raw data have been provided. For example, we might be interested in the one-sample t statistic for testing H_0: $\mu = 10$ versus one of the following, H_a: $\mu < 10$, H_a: $\mu > 10$, or H_a: $\mu \neq 10$ from a sample of $n = 23$ observations has a test statistic value of $t = 2.78$. Using software, find the exact P-value.

To obtain the *P*-value, follow these steps:

1. Enter the value of the test statistic into the SPSS for Windows Data Editor under the variable called ***teststat.***
2. From the SPSS for Windows main menu bar, click **Transform** and then click **Compute.** The "Compute Variable" window in Figure 16.2 appears.
3. In the "Target Variable" box, type in ***pvalue.***
4. In the "Functions" box, click ▼ until **CDF.T(q,df)** appears in the "Functions" box. Double-click on **CDF.T(q,df)** to move **CDF.T(?,?)** into the "Numeric Expression:" box. The **CDF.T(q,df)** function stands for the cumulative distribution function for the *t* distribution, and it calculates the area *to the left* of *q* under the correct *t* distribution.

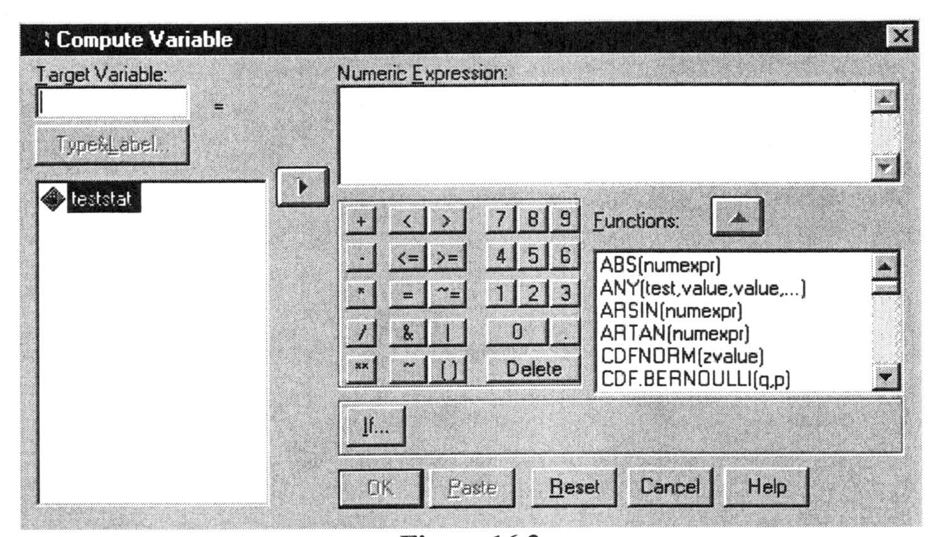

Figure 16.2

5. In the "Numeric Expression:" box, change the first **?** to ***teststat*** and the second **?** to **22** (23 – 1). For H_a: $\mu < 10$, **CDF.T(*teststat*,22)** should appear in the "Numeric Expression:" box. For H_a: $\mu > 10$, **1 – CDF.T(*teststat*,22)** should appear in the "Numeric Expression:" box. For H_a: $\mu \neq 10$, **2*(1 – CDF.T(ABS(*teststat*),22))** should appear in the "Numeric Expression:" box.
6. Click **OK.**

The *P*-value can be found in the SPSS for Windows Data Editor. By default, the number of decimal places for the variable ***pvalue*** is two. The number of digits after the decimal place can be changed in the variable view window. The *P*-value for H_a: $\mu < 10$ is 0.995. The *P*-value for H_a: $\mu > 10$ is 0.005. The *P*-value for H_a: $\mu \neq 10$ is 0.011.

Matched Pairs

We hear that listening to Mozart improves students' performance on tests. Perhaps pleasant odors have a similar effect. To test this idea, 21 subjects worked a paper-and-pencil maze while wearing a mask. See Example 16.3 in BPS. The mask was either unscented or carried a floral scent. The response variable is their average time in seconds on three trials. Each subject worked the maze with both masks, in a random order. The randomization is important because subjects tend to improve their times as they work a maze repeatedly. Table 16.1 in BPS gives the subjects' average times in seconds with both masks. Assess whether the floral scent significantly improved performance at the $\alpha = 0.05$.

This example is a **matched pairs study,** in which repeated measurements on the same subjects were obtained. These data can be retrieved from **ta16-01.por.** The unscented average time (***unscent***) is the first variable in the SPSS for Windows Data Editor and the scented average time (***scent***) is the second variable, where both variables are declared numeric 8.2. Before proceeding with the matched pairs *t* test, we must verify the assumption that the differences (the Difference column in Table 16.1, which equals ***unscent – scent***) come from a normal distribution. The first subject, for example, was 7.37 seconds slower wearing the scented mask, so the difference is negative. Because shorter times represent better performances, positive differences show that the subject did better when wearing the scented mask. If the differences have not been entered into the original SPSS for Windows data set, then a variable ***diff*** must be created to check the assumption of normality.

To create the variable ***diff,*** follow these steps:

1. Click **Transform,** and then click **Compute.** The "Compute Variable" window in Figure 7.3 appears, with the exception that ***unscent*** and ***scent*** appear in the window.
2. In the "Target Variable" box, type ***diff.***
3. Double-click ***unscent***, click the gray minus sign (–), and then double-click ***scent***. The expression ***unscent – scent*** appears in the "Numeric Expression:" box.
4. Click **OK.** The variable ***diff*** appears in the SPSS for Windows Data Editor.

Obtain a Q-Q plot and/or stemplot for the variable ***diff.*** A stemplot of the differences shows that the distribution is reasonably symmetric and appears reasonably normal in shape.

Set up H_0: $\mu = 0$ versus H_a: $\mu > 0$, where μ = the mean difference in the population from which the subjects were drawn. The null hypothesis says that no improvement occurs, and the alternative hypothesis says that unscented times are longer than scented times on the average.

To conduct a matched pairs *t* test, follow these steps:

1. Click **Analyze,** click **Compare Means,** and then click **Paired–Samples T Test.** The "Paired-Samples T Test" window in Figure 16.3 appears.

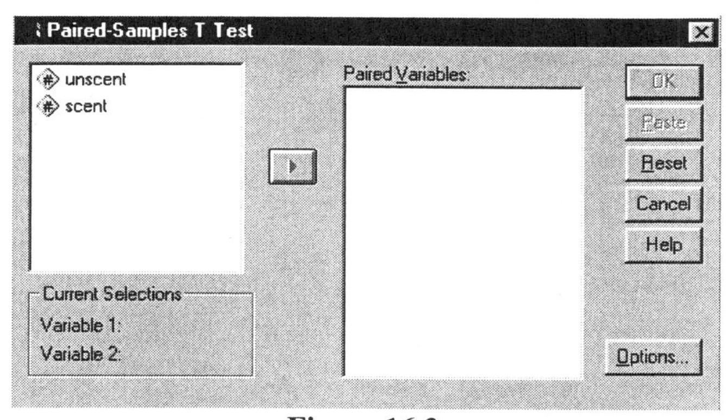

Figure 16.3

2. Click ***unscent.*** The variable ***unscent*** appears after "Variable 1" in the "Current Selections" box.
3. Click ***scent.*** The variable ***scent*** appears after "Variable 2" in the "Current Selections" box.
4. Click ▸. The expression ***unscent – scent*** appears in the "Paired Variables" box.

5. By default, a 95% confidence interval for μ will be part of the matched pairs *t* test output. To change the confidence level, click **Options,** change 95 in the "Confidence Interval" box to the desired confidence level.
6. Click **Continue.**
7. Click **OK.**

Table 16.3 is part of the resulting SPSS for Windows output.

Paired Samples Test

	Paired Differences					t	df	Sig. (2-tailed)
				95% Confidence Interval of the Difference				
	Mean	Std. Deviation	Std. Error Mean	Lower	Upper			
Pair 1 UNSCENT - SCENT	.9567	12.5479	2.7382	-4.7551	6.6684	.349	20	.730

Table 16.3

Note: The same results would have been obtained if we had applied the one-sample *t* test to the variable ***diff*** using a test value of 0 (follow the directions given above).

Because of the one-sided alternative, we are interested only in the upper right tail above 0.349. Therefore, the *P*-value is 0.365. This is obtained by taking the value under the "Sig. (2-tailed)" column and dividing it by 2 (0.730/2).

The data do not support the claim that floral scents improve performance. The average improvement is small: just 0.9567 seconds. This small improvement is not statistically significant at the 5% level. In addition, we are 95% confident that the mean difference in the population from which the subjects were drawn lies somewhere between –4.7551 and 6.6684 seconds.

Chapter 17. Two-Sample Problems

Topics covered in this chapter:

- **Two-Sample *t* Procedures**
- **The *F* Test for Equality of Variances**

This chapter introduces the use of the *t* distribution in inferential statistics for comparing two means. The two-sample problem examined in this section compares the responses in the two groups, where the responses in each group are independent of those in the other group. Assuming the two samples come from normal populations, the **two-sample *t* procedure** is the correct test to apply. It is also of interest to ask if the variances differ from one group to the next, and this can be tested using the ***F* test for equality of variances.**

Two-Sample *t* Procedures

How quickly do synthetic fabrics, such as polyester, decay in landfills? A researcher buried polyester strips in the soil for different lengths of time, then dug up the strips and measured the force required to break them. Breaking strength is easy to measure and is a good indicator of decay. Lower strength means the fabric has decayed.

In part of the study, the researcher buried 10 polyester strips in well-drained soil in the summer. Five of the strips, chosen at random, were dug up after 2 weeks; the other 5 were dug up after 16 weeks. Here are the breaking strengths in pounds:

2 weeks: 118, 126, 126, 120, 129
16 weeks: 124, 98, 110, 140, 110.

Is there sufficient evidence at the 0.10 level to conclude that the polyester decays more in 16 weeks than in 2 weeks?

Before proceeding with the two-sample *t* test, we must verify that the assumption of normally distributed data in both groups is reasonably satisfied. Both populations are reasonably normal (which can be determined using stemplots and/or Q-Q plots for the two groups of data), as far as we can tell from so few observations.

Set up H_0: $\mu_1 = \mu_2$ versus H_a: $\mu_1 > \mu_2$, where μ_1 = the true mean breaking strength in the entire population of polyester fabric buried for 2 weeks and μ_2 = the true mean breaking strength in the entire population of polyester fabric buried for 16 weeks. Recall that less breaking strength implies more decay.

Retrieve the data from **ex17-02.por.** The SPSS for Windows Data Editor contains the variables ***weeks*** and ***strength*** (both declared as numeric 11.0). The values for the variable ***weeks*** are 2 (for the 2 weeks group) and 16 (for the 16 weeks group).

To perform the two-sample *t* test, follow these steps:

1. Click **Analyze,** click **Compare Means,** and then click **Independent-Samples T Test.** The "Independent-Samples T Test" window in Figure 17.1 appears.
2. Click ***strength,*** then click ▸ to move ***strength*** into the "Test Variable(s)" box.

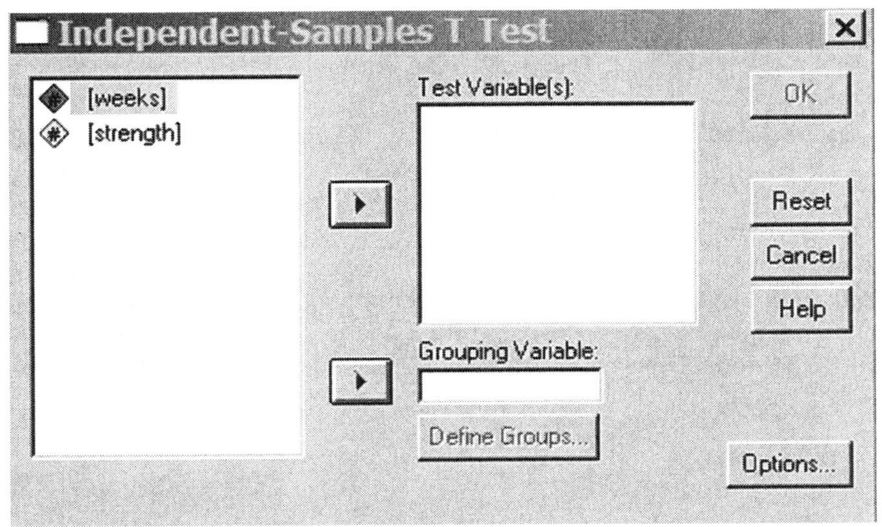

Figure 17.1

3. Click ***weeks,*** then click ▸ to move ***weeks*** into the "Grouping Variable" box.
4. Click **Define Groups.** The "Define Groups" window in Figure 17.2 appears.

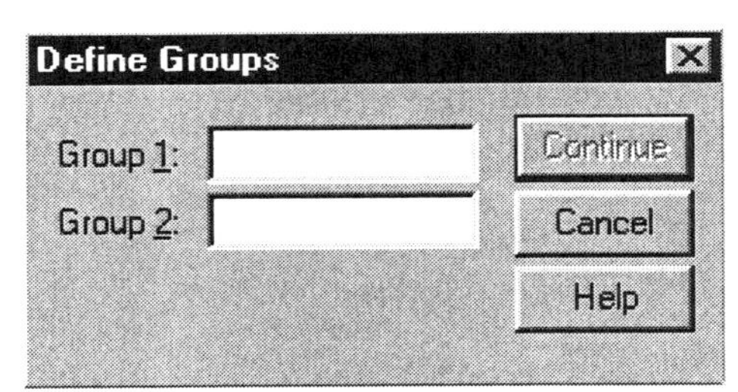

Figure 17.2

5. Type ***2*** in the "Group 1" box. Press the **Tab** key. Type ***16*** in the "Group 2" box. *Note:* The group values must be typed in exactly as they appear in the SPSS for Windows Data Editor. Click **Continue. Weeks(2 16)** appears in the "Grouping Variable" box.
6. By default, a 95% confidence interval for $\mu_1 - \mu_2$ (the difference in population means) will be part of the two-sample *t* test output. To change the confidence level, click **Options** and change 95 in the "Confidence Interval" box to the desired confidence level (for instance, **90**).
7. Click **Continue.**
8. Click **OK.**

Table 17.1 is part of the resulting SPSS for Windows output.

Independent Samples Test

	Levene's Test for Equality of Variances		t-test for Equality of Means					90% Confidence Interval of the Difference	
	F	Sig.	t	df	Sig. (2-tailed)	Mean Difference	Std. Error Difference	Lower	Upper
Equal variances assumed	5.581	.046	.989	8	.352	7.40	7.483	-6.516	21.316
Equal variances not assumed			.989	4.651	.371	7.40	7.483	-7.933	22.733

Table 17.1

The *F* Test for Equality of Variances

SPSS for Windows reports the results of two t procedures: the pooled two-sample t procedure (assumes equal population variances) and a general two-sample t procedure (does not assume equal population variances). To determine which t procedure to use, SPSS for Windows performs Levene's Test for Equality of Variances for H_0: $\sigma_1^2 = \sigma_2^2$ versus H_a: $\sigma_1^2 \neq \sigma_2^2$. The F test statistic for the Levene's test is obtained by computing a one-way analysis of variance (see Chapter 22) on the absolute deviations of each case from its group mean. The P-value for Levene's test of 0.046 is located under the "Sig." column. We reject H_0 in favor of H_a; that is, there is sufficient evidence at the 0.10 level to conclude that the population variances are unequal. Thus, we use the two-sample t procedure for which equal variances are not assumed.

Because of the one-sided alternative, we are interested only in the upper right tail above 0.989. Therefore, the P-value is 0.1855. This is obtained by taking the value under the "Sig. (2-tailed)" column for the "Equal variances not assumed" case and dividing it by 2 (0.371/2).

We fail to reject H_0; that is, the experiment did not find convincing evidence that polyester decays more in 16 weeks than 2 weeks. In addition, we are 90% confident that the mean strength change between 2 and 16 weeks, $\mu_1 - \mu_2$, lies somewhere between −7.9329 and 22.7329 pounds.

Note: There is a slight discrepancy in the confidence interval bounds between BPS and SPSS for Windows. The critical value t^* in the BPS example is based on 4 degrees of freedom, whereas the critical value t^* in SPSS for Windows is based on 4.651 degrees of freedom.

Chapter 18. Inference about a Population Proportion

Topics covered in this chapter:

- **Confidence Intervals for a Single Proportion**
- **Significance Tests for a Proportion**

Chapter 18 in BPS focuses on inference about **single population proportions.** Although in the professional version of SPSS for Windows, programs could be written to perform inference for proportions, this option is not available in the Student Version. The computations discussed in this chapter can be accomplished using a calculator or another statistical package.

After the test statistic is computed for a test of significance, SPSS for Windows can be used to compute the *P*-value for the test as described in Chapter 14 of this manual. Note that the problems in this chapter use the *z* distribution. As a result, the *P*-values would be computed using the function **CDF.NORMAL(q,0,1).**

Chapter 14 of this manual shows you how to calculate the z statistic and the probability associated with it.

To test your knowledge of SPSS and your understanding of Chapter 18 in BPS, challenge yourself to replicate Examples 18.3, 18.4 and 18.5 in BPS about risky behavior. Then replicate the "Is this coin Fair?" Examples 18.8 and 18.9. Finally, open the data for Exercise 14.48 **(ex14-48.por)** and complete the same calculations.

Chapter 19. Comparing Two Proportions

Topics covered in this chapter:

- **Confidence Intervals for Comparing Proportions**
- **More Accurate Confidence Intervals**
- **Significance Tests for Comparing Proportions**

Chapter 19 in BPS focuses on inference about **two population proportions.** Although in the professional version of SPSS for Windows, programs could be written to perform inference for proportions, this option is not available in the Student Version. The computations discussed in this chapter can be accomplished using a calculator or another statistical package.

As was described in Chapter 18 of this manual, you will be computing z statistics and probabilities. Chapter 14 of this manual shows you how to calculate both the z statistic and the probability associated with it.

To test your knowledge of SPSS and your understanding of Chapter 19 in BPS, challenge yourself to replicate Examples 19.2 and 19.3 in BPS about the value of preschool in reducing later utilization of social services. Then replicate the calculations in "Choosing a mate," Examples 19.4 and 19.5. Finally, retrieve the data for the EESE story "Radar Detection and Speeding" from the CD-ROM and complete the same calculations.

Chapter 20. Two Categorical Variables: The Chi-Square Test

Topics covered in this chapter:

- **Two-way Tables**
- **The Chi-Square Test of Independence**

This chapter introduces the notion of analyzing two categorical variables. The data are often summarized using a **two-way table. Inferential** procedures are applied to test whether the two categorical variables are independent.

Two-way Tables

Cocaine addicts need the drug to feel pleasure. Perhaps giving them a medication that fights depression will help them stay off cocaine. A three-year study compared an antidepressant called desipramine with lithium (a standard treatment for cocaine addiction) and a placebo. The subjects were 72 chronic users of cocaine who wanted to break their drug habit. Twenty-four of the subjects were randomly assigned to each treatment. The data are displayed in a 3 × 2 contingency table (Table 20.1). The question of interest is whether these data give good evidence that the proportions of successes for these three treatments differ in the population of all cocaine addicts.

Table 20.1 Counts of Relapse by Treatment

Observed Counts for Relapses			
	Relapse		
Treatment	No	Yes	Total
Desipramine	14	10	24
Lithium	6	18	24
Placebo	4	20	24
Total	24	48	72

The entries in this table are the observed, or sample, counts. For example, 14 cocaine addicts received the treatment desipramine and did not experience a relapse in cocaine usage. Note that the marginal totals are given with the table. They are not part of the raw data but are calculated by summing over the rows or columns. The row totals are the numbers of observations sampled in the three populations. The grand total, 72, can be obtained by summing the row or the column totals. It is the total number of observations (cocaine addicts) in the study.

Note: These data were entered using the three variables of ***treatment***, ***relapse***, and ***weight*** where ***weight*** represents the count for a particular combination of ***treatment*** and ***relapse.*** Prior to any analyses, the **Weight Cases** option under **Data** was activated.

To describe the relationship between these two categorical variables, you can generate and compare percentages. Each cell count can be expressed as a percentage of the grand total, the row total, and the column total. To generate these percentages, follow these steps:

1. Click **Analyze,** click **Descriptive statistics,** and then click **Crosstabs.** The "Crosstabs" window in Figure 20.1 appears.

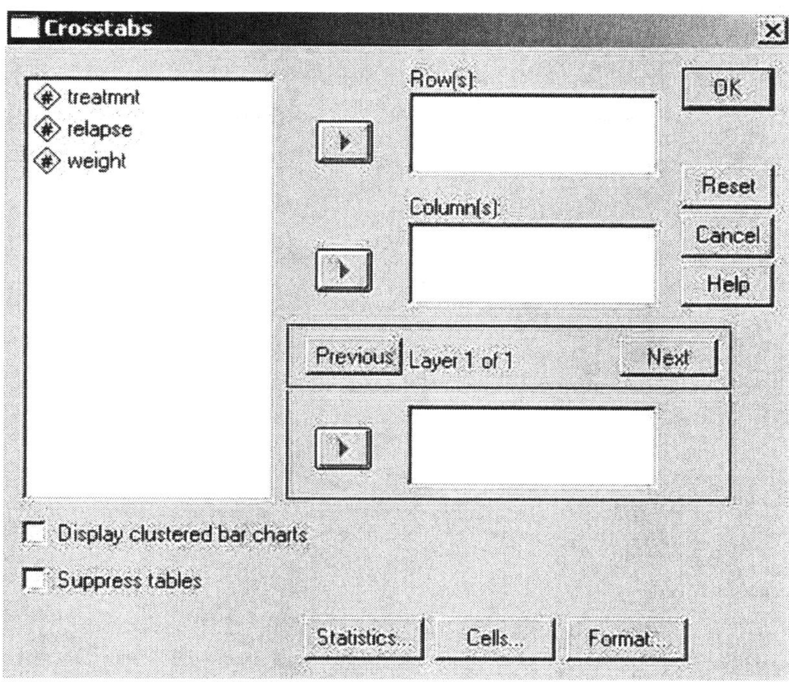

Figure 20.1

2. Click ***treatmnt,*** then click ▸ to move ***treatmnt*** into the "Row(s)" box.
3. Click ***relapse,*** then click ▸ to move ***relapse*** into the "Column(s)" box.
4. Click the **Cells** button. The "Crosstabs: Cell Display" window in Figure 20.2 appears.

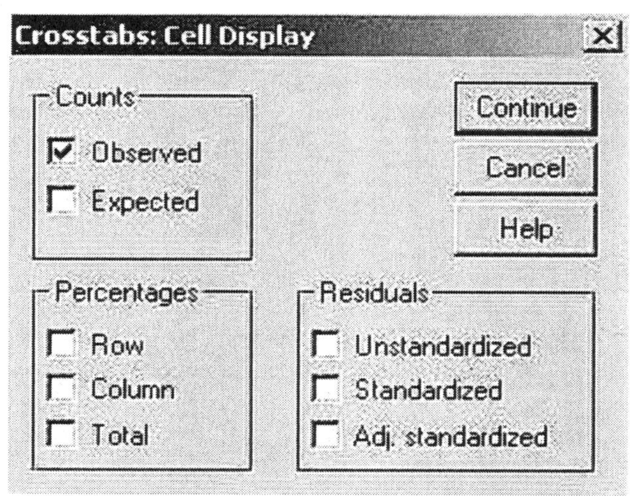

Figure 20.2

5. Click **Row, Column,** and **Total** within the "Percentages" box so a check mark (✓) appears before each type of percentage.
6. Click **Continue.**
7. Click **OK.**

The resulting SPSS for Windows output is shown in Table 20.2.

TREATMNT * RELAPSE Crosstabulation

			RELAPSE		
			No	Yes	Total
TREATMNT	Desipramine	Count	14	10	24
		% within TREATMNT	58.3%	41.7%	100.0%
		% of Total	19.4%	13.9%	33.3%
	Lithium	Count	6	18	24
		% within TREATMNT	25.0%	75.0%	100.0%
		% of Total	8.3%	25.0%	33.3%
	Placebo	Count	4	20	24
		% within TREATMNT	16.7%	83.3%	100.0%
		% of Total	5.6%	27.8%	33.3%
Total		Count	24	48	72
		% within TREATMNT	33.3%	66.7%	100.0%
		% of Total	33.3%	66.7%	100.0%

Table 20.2

Each cell contains four entries, which are labeled at the beginning of each row. The "Count" is the cell count. The "% within TREATMNT" is the cell count expressed as a percentage of the row total. As an example, out of all 24 individuals who received desipramine, 14, or 58.3%, did not relapse. The "% within RELAPSE" is the cell count expressed as a percentage of the column total. For instance, out of the 48 individuals who did relapse, 10, or 20.8%, were treated with desipramine. The "% of Total" is the cell count expressed as a percentage of the grand total. Of the 72 total individuals who participated in the study, 4, or 5.6%, received a placebo and did not relapse.

In this example, most interest lies in the effect of treatment on the distribution of relapses. Thus, to compare the treatments, the row percentages or the percentages within treatments are examined. A higher percentage of individuals receiving desipramine did not relapse when compared to both those receiving lithium and those receiving the placebo. Also, a higher percentage of those individuals receiving the placebo experienced a relapse when compared to those receiving lithium and desipramine. This suggests that desipramine does much better than placebo in preventing relapses, with lithium in between.

The Chi-Square Test of Independence

By comparing the sample proportion of successes, we can describe the differences among the three treatments for cocaine addiction. The chi-square test of independence, which compares the observed and expected counts, can be used to assess the extent to which these differences can be plausibly attributed to chance. The null and alternative hypotheses are H_0: there is no association between treatment and relapse and H_a: there is an association between treatment and relapse.

To perform the χ^2 test of independence, follow these steps:

1. Click **Analyze,** click **Descriptive statistics,** and then click **Crosstabs.**
2. Click ***treatmnt,*** then click ▸ to move ***treatmnt*** into the "Row(s)" box.
3. Click ***relapse,*** then click ▸ to move ***relapse*** into the "Column(s)" box.
4. If you are interested in including the expected cell counts within the contingency table that will result from the analysis, click the **Cells** button and click **Expected** under **Counts** so a check mark (✔) appears before **Expected.** If a check marks appear before **Row, Column,** and **Total** under the "Percentages" box, click on each of these terms so the check marks disappear. Then click **Continue.**

5. Click the **Statistics** button.
6. Click **Chi-Square.**
7. Click **Continue.**
8. Click **OK.**

The resulting SPSS for Windows output is shown in Tables 20.3 and 20.4.

In comparing the expected and observed counts, the treatment desipramine had more successes and fewer failures than we would expect if all three treatments had the same success rate in the population. The other two treatments had fewer successes and more failures than expected, with the differences more pronounced with those receiving placebo. All of the expected cell counts are moderately large, so the χ^2 distribution should reasonably approximate the *P*-value. The test statistic value (Pearson Chi-square) is $\chi^{2*} = 10.500$ with $df = 2$, and *P*-value (Asymptotic Significance) = 0.005. The Chi-square test confirms that the data contain clear evidence against the null hypothesis that three treatments have the same distribution of relapses. Under H_o, the chance of obtaining a value of χ^2 greater than or equal to the calculated value of 10.500 is very small, 0.005.

TREATMNT * RELAPSE Crosstabulation

			RELAPSE		
			No	Yes	Total
TREATMNT	Desipramine	Count	14	10	24
		Expected Count	8.0	16.0	24.0
	Lithium	Count	6	18	24
		Expected Count	8.0	16.0	24.0
	Placebo	Count	4	20	24
		Expected Count	8.0	16.0	24.0
Total		Count	24	48	72
		Expected Count	24.0	48.0	72.0

Table 20.3

Chi-Square Tests

	Value	df	Asymp. Sig. (2-sided)
Pearson Chi-Square	10.500[a]	2	.005
Likelihood Ratio	10.438	2	.005
Linear-by-Linear Association	9.245	1	.002
N of Valid Cases	72		

a. 0 cells (.0%) have expected count less than 5. The minimum expected count is 8.00.

Table 20.4

Chapter 21. Inference for Regression

Topic covered in this chapter:

- **Simple Linear Regression**

The descriptive analysis discussed in Chapter 5 for relationships between two quantitative variables leads to formal inference. This chapter focuses on demonstrating how SPSS for Windows can be used to perform inference for **simple linear regression.**

Simple Linear Regression

Example 21.1 in BPS discusses the fact that infants who cry easily may be more easily stimulated than others and this may be a sign of higher IQ. Child development researchers explored the relationship between the crying of infants 4 to 10 days old and their later IQ test scores. A snap of a rubber band on the sole of the foot caused the infants to cry. The researchers recorded the crying and measured its intensity by the number of peaks in the most active 20 seconds. They later measured the children's IQ at age 3 using the Stanford-Binet IQ test. Table 21.1 contains data on 38 infants.

Crying	IQ	Crying	IQ	Crying	IQ	Crying	IQ
10	87	20	90	17	94	12	94
12	97	16	100	19	103	12	103
9	103	23	103	13	104	14	106
16	106	27	108	18	109	10	109
18	109	15	112	18	112	23	113
15	114	21	114	16	118	9	119
12	119	12	120	19	120	16	124
20	132	15	133	22	135	31	135
16	136	17	141	30	155	22	157
33	159	13	162				

Table 21.1

These data can be opened form the SPSS Data Editor Window by clicking on **ta21-01.por.**

The response variable is the IQ test score and is plotted on the y axis. The explanatory variable is the count of crying peaks and is plotted on the x axis. Figure 21.1 is the scatterplot of the data set with the fitted line superimposed. There is a moderate positive linear relationship, with no extreme outliers or potentially influential observations. As a result, it is now of interest to model this relationship.

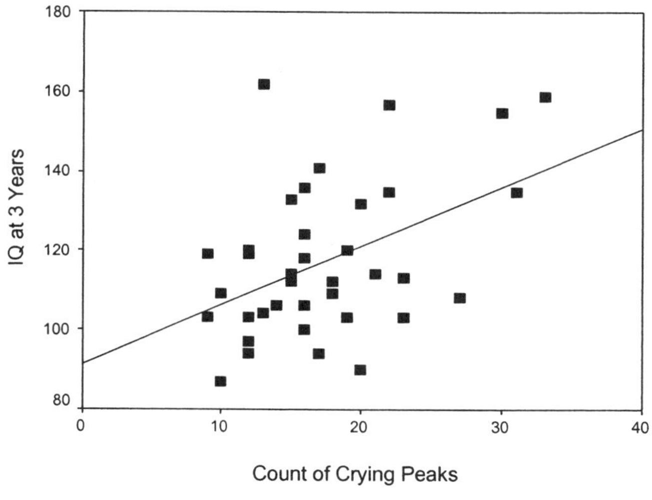

Figure 21.1

To obtain the correlation and the least-squares regression line, follow these steps:

1. Click **Analyze,** click **Regression,** and then click **Linear.** The "Linear Regression" window in Figure 21.2 appears.

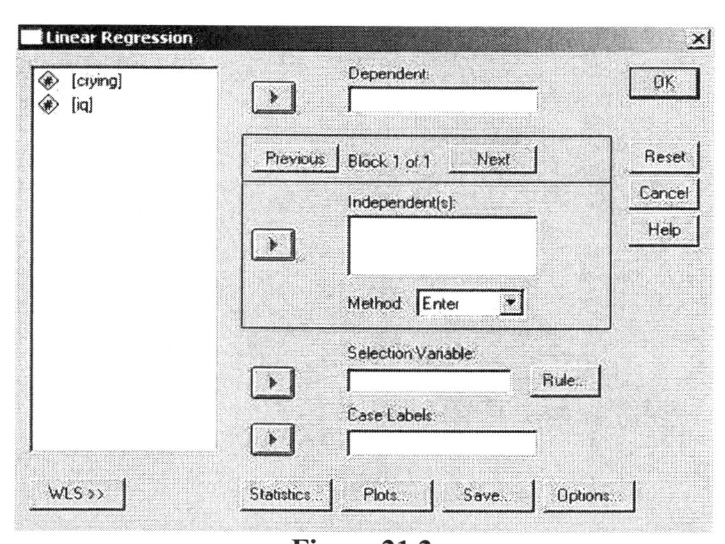

Figure 21.2

2. Click ***iq,*** then click ‣ to move ***iq*** into the "Independent(s)" box.
3. Click ***crying,*** then click ‣ to move ***crying*** into the "Dependent" box.
4. If you are interested in a normal probability plot for the residuals, click the **Plots** box (Figure 21.3 appears). Select the **Normal probability plot** option in the "Standardized Residual Plots" box.
5. Click **Continue.**

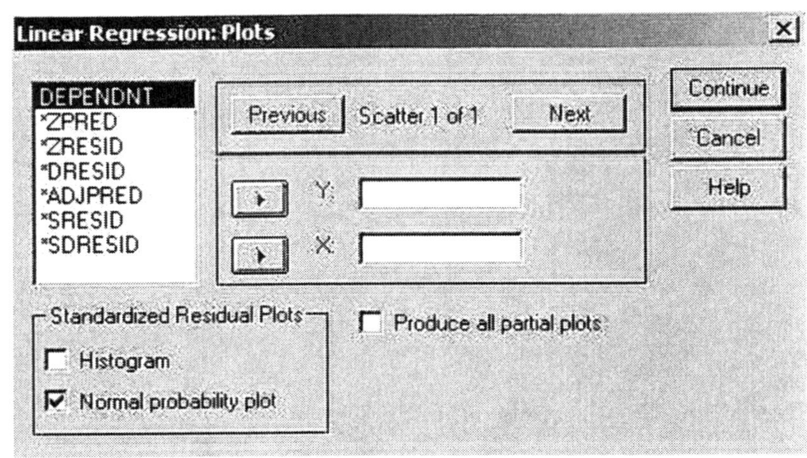

Figure 21.3

6. If you are interested in saving predicted values, residuals, distance measures, influential statistics, and prediction intervals, click the **Save** box (Figure 21.4 appears). Check the desired options and click **Continue.**

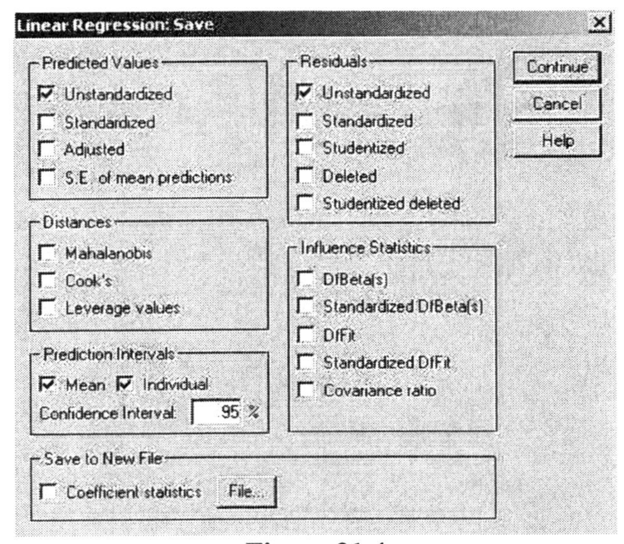

Figure 21.4

7. If you are interested in obtaining the confidence interval for the slope, click the **Statistics** box (Figure 21.5 appears). Check **Confidence intervals** in the "Regression Coefficients" box.
8. Click **Continue.**
9. Click **OK.**

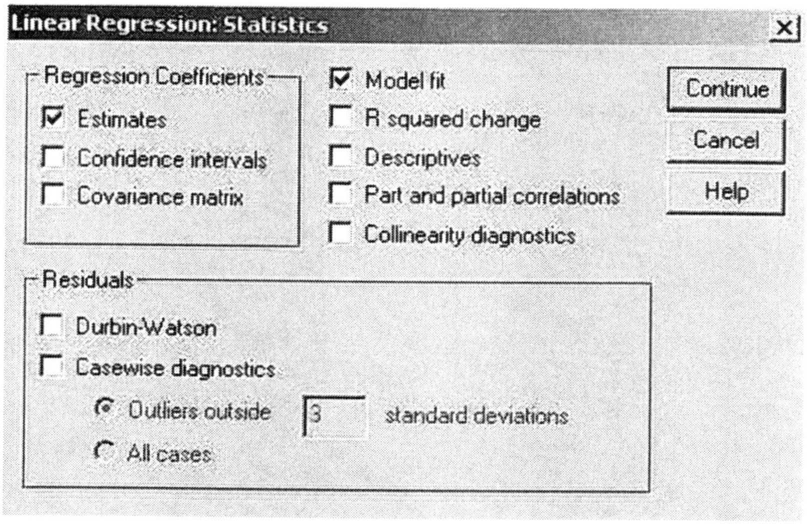

Figure 21.5

Tables 21.2 and 21.3 and are part of the resulting SPSS for Windows output.

Model Summary[b]

Model	R	R Square	Adjusted R Square	Std. Error of the Estimate
1	.455[a]	.207	.185	17.499

a. Predictors: (Constant),

b. Dependent Variable:

Table 21.2

Coefficients[a]

Model		Unstandardized Coefficients		Standardized Coefficients	t	Sig.	95% Confidence Interval for B	
		B	Std. Error	Beta			Lower Bound	Upper Bound
1	(Constant)	91.268	8.934		10.216	.000	73.149	109.388
		1.493	.487	.455	3.065	.004	.505	2.481

a. Dependent Variable:

Table 21.3

Tables 21.2 and 21.3 show the correlation, the y-intercept, and the slope. The correlation of $r = 0.455$ indicates that a strong, positive, linear association exists between IQ and Counts of crying peaks. Recall that the value for the Multiple R found in Table 21.2 gives the absolute value (or magnitude) of r. If you are interested in obtaining the actual value of r you must also look at the slope to determine whether the value of r is positive or negative. In addition, the equation of the least squares regression line is:

$$\hat{iq} = 91.27 + 1.493\ Crying$$

The estimated standard deviation of the model (often referred to as the MSE) is 17.499.

Now that a fitted model has been developed, the residuals of the model should be examined. Figure 21.6 is a normal probability plot of the standardized residuals. Because the plotted values are fairly linear, the assumption of normality seems reasonable.

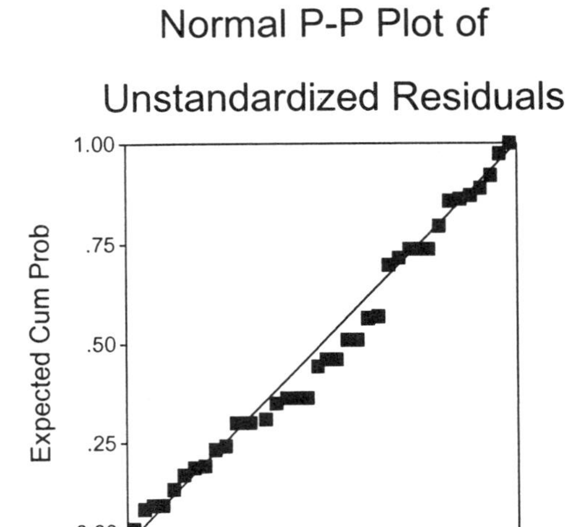

Figure 21.6

In addition, we can examine whether the residuals display any systematic pattern when plotted against other variables. Recall that the unstandardized residuals (and various other values) were saved into the Data Editor in Step 6. Figure 21.7 displays the Data Editor containing the values that were saved, and the variable ***res_1*** contains the residuals. Figure 21.8 is a scatterplot of ***res_1*** and ***crying.*** No unusual patterns or values are observed.

Untitled - SPSS Data Editor

File Edit View Data Transform Analyze Graphs Utilities Window Help

1 : crying 10

	crying	iq	pre_1	res_1	lmci_1	umci_1	lici_1	uici_1
1	10	87	13.20220	-3.20220	9.92013	16.48426	1.89960	24.50480
2	20	90	13.61817	6.38183	10.56513	16.67120	2.37994	24.85640
3	17	94	14.17279	2.82721	11.41199	16.93360	3.01041	25.33518
4	12	94	14.17279	-2.17279	11.41199	16.93360	3.01041	25.33518
5	12	97	14.58876	-2.58876	12.03444	17.14309	3.47565	25.70188
6	16	100	15.00473	.99527	12.64284	17.36663	3.93426	26.07520
7	19	103	15.42070	3.57930	13.23348	17.60793	4.38618	26.45523
8	12	103	15.42070	-3.42070	13.23348	17.60793	4.38618	26.45523
9	9	103	15.42070	-6.42070	13.23348	17.60793	4.38618	26.45523
10	23	103	15.42070	7.57930	13.23348	17.60793	4.38618	26.45523
11	13	104	15.55936	-2.55936	13.42564	17.69308	4.53532	26.58341

Figure 21.7

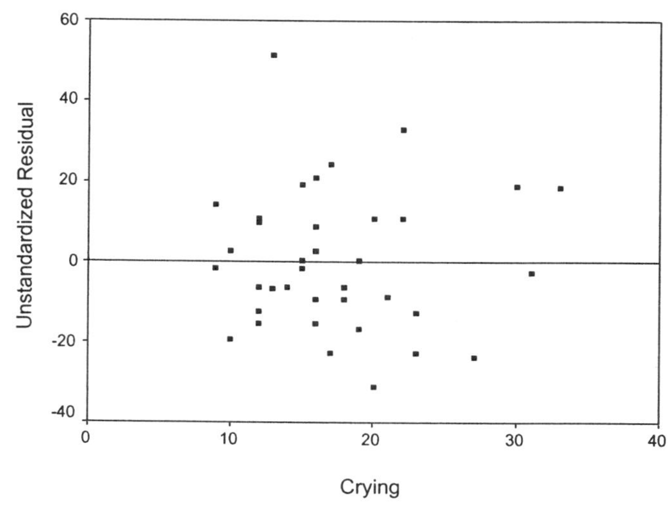

Figure 21.8

The hypotheses appropriate for testing this conjecture are H_0: $\beta = 0$ says that crying has no straight-line relationship with IQ and H_a: $\beta > 0$. According to Table 21.3, the computed test statistic $t = 3.07$ and the *P*-value 0.004. There is strong evidence that there is a straight-line relationship between IQ and crying. [Note that some software allows you to choose between one-sided and two-sided alternatives in tests of significance. Other software always reports the *P*-value for the two-sided alternative. If your alternative hypothesis is one-sided, you must divide *P* by 2]. Table 21.3 also shows that a 95% confidence interval for β is (0.505, 2.481).

It is also of interest to construct confidence intervals for the mean response and prediction intervals for a future observation. Figure 21.7 displays 95% confidence intervals for the mean response for each observation in the data set and 95% prediction intervals for future observations equal to each observation in the data set. The variables (***lmci_1***, ***umci_1***) represent the 95% confidence interval, and the variables (***lici_1***, ***uici_1***) represent the 95% prediction interval.

Chapter 22. One-Way Analysis of Variance: Comparing Several Means

Topic covered in this chapter:

- **One-Way ANOVA**

This chapter describes how to perform **one-way ANOVA** using SPSS for Windows for determining whether the means from several populations differ. Specifically, the null and alternative hypotheses for one-way ANOVA are H_0: $\mu_1 = \mu_2 = ... = \mu_I$ and H_a: not all of the μ_i are equal.

One-Way ANOVA

A study of highway gas mileage for three populations of vehicles compared midsize cars, pickups, and sport utility vehicles (SUVs). The highway gas mileage of 31 midsize cars, 31 SUVs, and 14 pickup trucks are given in Table 22.1 in BPS. This table can be read into the SPSS Data Editor for Windows by opening the file **ta22-01.por.** Note that when this data set is opened, the variable TYPE must be numerically coded (labels may also be used). In this example, codes and labels are as follows: 1 = midsize cars, 2 = SUVs, and 3 = pickup trucks.

Before proceeding with ANOVA, we must verify that the assumptions of (1) normally distributed data and (2) equality of standard deviations in the various groups are reasonably satisfied. Note that to determine whether the equality of standard deviations (or variances) is reasonably satisfied, SPSS for Windows performs Levene's Test for Equality of Variances. Because the data appear reasonably normal (which can be demonstrated using normal quantile plots for the three groups of data) and the assumption of equal standard deviations is reasonably satisfied, we can proceed with ANOVA. Specifically, the null and alternative hypotheses are H_0: $\mu_1 = \mu_2 = \mu_3$ and H_a: not all of the μ_i are equal.

To perform a one-way ANOVA with post hoc multiple comparisons, follow these steps:

1. Click **Analyze,** click **Compare Means,** and then click **One-Way ANOVA.** The "One-Way ANOVA" window in Figure 22.1 appears (see next page).
2. Click ***mpg,*** then click ‣ to move ***mpg*** into the "Dependent List" box.
3. Click ***type,*** then click ‣ to move ***type*** into the "Factor" box.
4. If you are interested in multiple comparisons, you can click on the **Post Hoc** box. The SPSS window in Figure 22.2 appears (see next page). Click on the desired multiple comparison test procedure (e.g., LSD).
5. Click **Continue.**

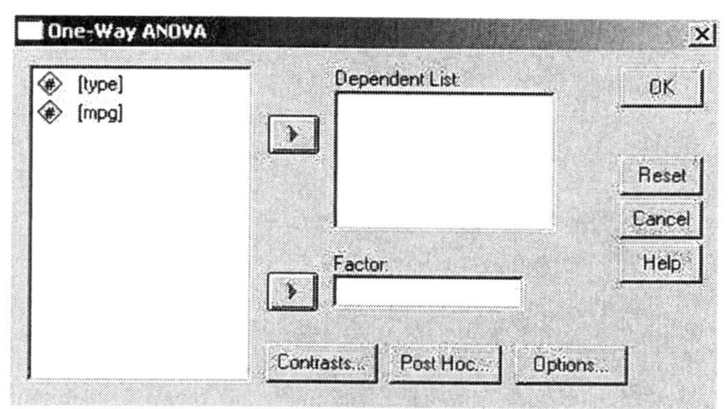

Figure 22.1

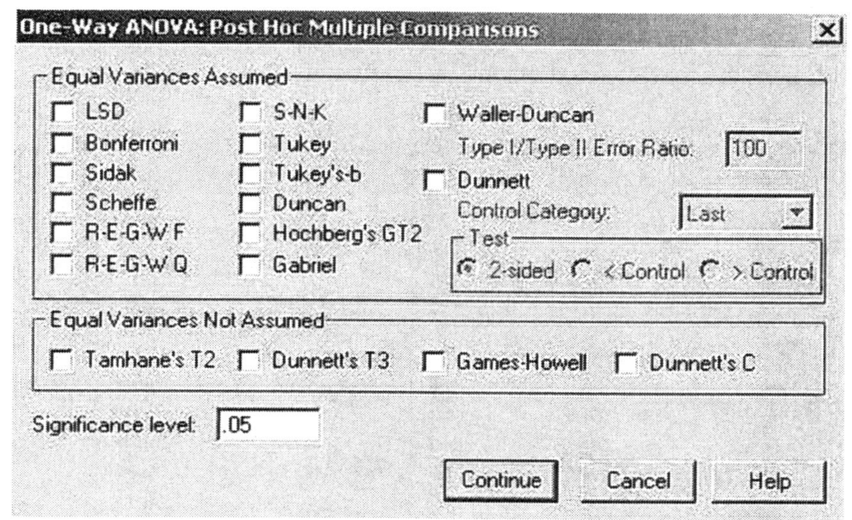

Figure 22.2

6. If you are interested in either descriptive statistics, a means plot, or Levene's Test for the Equality of Variances, you can click on the **Options** box. The "One-Way ANOVA: Options" window in Figure 10.3 appears. If you are interested in descriptive statistics, click **Descriptive.** If you are interested in a means plot, click **Means plot.** If you are interested in performing Levene's Test, click **Homogeneity-of-variance.**
7. Click **Continue**.
8. Click **OK.**

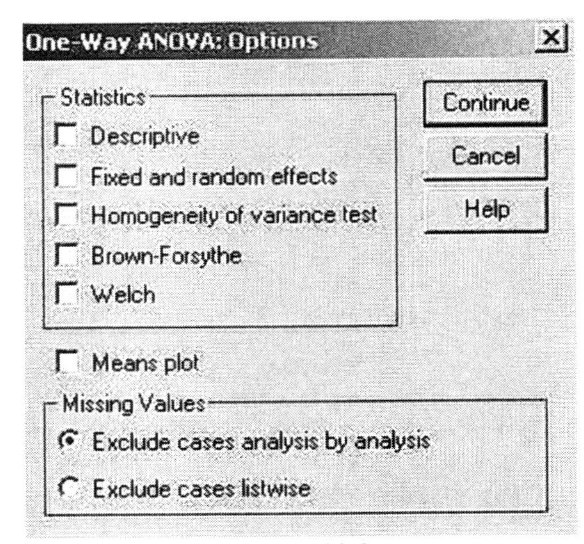

Figure 22.3

For illustration purposes, SPSS for windows output for the Descriptives table (Table 22.1), ANOVA table (Table 22.1), the LSD method for multiple comparison (Table 22.3), and a means plot (Figure 22.4) have been provided.

Descriptives

	N	Mean	Std. Deviation	Std. Error	95% Confidence Interval for Mean		Minimum	Maximum
					Lower Bound	Upper Bound		
1	31	27.90	2.561	.460	26.96	28.84	23	33
2	31	22.68	3.673	.660	21.33	24.02	17	29
3	14	21.29	2.758	.737	19.69	22.88	17	26
Total	76	24.55	4.174	.479	23.60	25.51	17	33

Table 22.1

ANOVA

	Sum of Squares	df	Mean Square	F	Sig.
Between Groups	606.448	2	303.224	31.607	.000
Within Groups	700.341	73	9.594		
Total	1306.789	75			

Table 22.2

Because the *P*-value is less than 0.001 (from Table 22.2), there is sufficient evidence to reject H_0 in favor of H_a. This indicates that some μ_i differ. An initial comparison for the gas mileage of the three types of vehicles can be made from the descriptive statistics in Table 22.1.

In many studies specific questions cannot be formulated in advance. If H_0 is rejected, we would like to know which means differ. One procedure that is commonly used for multiple comparisons is the least-significant differences (LSD) method. The output of the LSD method, found in Table 22.3, indicates significantly different means with an asterisk (*). The LSD method indicates that the means for groups 1 and 2 and 1 and 3 differ significantly. Groups 2 and 3 are not significantly different. In addition, the means plot, in Figure 22.4, is useful in demonstrating the relationships between the various means.

Multiple Comparisons

Dependent Variable:
LSD

(I)	(J)	Mean Difference (I-J)	Std. Error	Sig.	95% Confidence Interval	
					Lower Bound	Upper Bound
1	2	5.23*	.787	.000	3.66	6.79
	3	6.62*	.997	.000	4.63	8.61
2	1	-5.23*	.787	.000	-6.79	-3.66
	3	1.39	.997	.167	-.60	3.38
3	1	-6.62*	.997	.000	-8.61	-4.63
	2	-1.39	.997	.167	-3.38	.60

*. The mean difference is significant at the .05 level.

Table 22.3

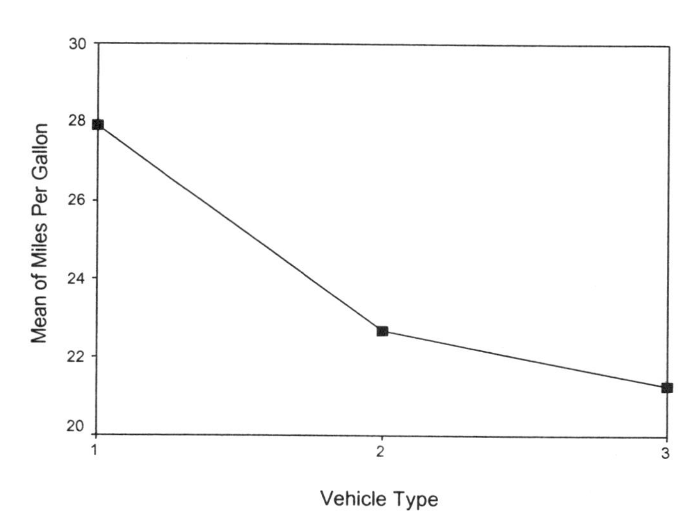

Figure 22.4

Chapter 23. Nonparametric Tests

Topics covered in this chapter:

- **Wilcoxon Rank Sum Test**
- **Wilcoxon Signed Rank Test**
- **Kruskal-Wallis Test**

This chapter introduces one class of **nonparametric** procedures, tests that can replace *t* tests and one-way analysis of variance when the normality assumption for those tests is not met. When distributions are strongly skewed, the mean may not be the preferred measure of center. The focus of these tests is on medians rather than means. All three of these tests use ranks of the observations in calculating the test statistic.

Wilcoxon Rank Sum Test

The **Wilcoxon rank sum test** is the nonparametric counterpart of the parametric independent *t* test. It is applied to situations in which the normality assumption underlying the parametric independent *t* test has been violated or questionably met. The focus of this test is on medians rather than means. An alternate form of this test, the one used by SPSS for Windows, is the Mann-Whitney *U* test.

Food sold at outdoor fairs and festivals may be less safe than food sold in restaurants because it is prepared in temporary locations and often by volunteer help. What do people who attend fairs think about the safety of food served? One study asked this question of people at a number of fairs in the Midwest:

How often do you think people become sick because of food they consume prepared at outdoor fairs and festivals? The possible responses were: 1 = very rarely, 2 = once in a while, 3 = often, 4 = more often than not, and 5 = always.

In all, 303 people answered the question. Of these, 196 were women and 107 were men. Is there good evidence that men and women differ in their perceptions about food safety at fairs? The data are presented in Table 23.1 as a two-way table of counts.

	Responses					
Gender	1	2	3	4	5	Total
Female	13	108	50	23	2	196
Male	22	57	22	5	1	107
Total	35	165	72	28	3	303

Table 23.1 Responses by Gender

We would like to know whether men or women are more concerned about food safety. Whereas a Chi-square test could be applied to answer the general question, this test ignores the ordering of the responses and so does not use all of the available information. Because the data are ordinal, a test based on ranks makes sense. One can use the Wilcoxon rank sum test for the hypotheses:

H_0: men and women do not differ in their responses
H_a: one of the two genders gives systematically larger responses than the other.

The data were entered into SPSS for Windows using 10 rows and 3 columns with the variable names gender (declared numeric 8.0 with value labels 1 = Female and 2 = Male), sick (declared numeric 8.0 with value labels 1 = very rarely, 2 = once in a while, 3 = often, 4 = more often than not, and 5 = always), and weight (declared numeric 8.0), where weight represents the count of individuals for each gender who selected each of the five response options.

It is important to note that the grouping variable (in this case gender) must be a numeric variable (not entered as M's and F's). Also, before performing analyses, the weighting option under Data and Weight Cases was activated.

To conduct a Wilcoxon rank sum test (or Mann-Whitney *U* test), follow these steps:

1. Click **Analyze,** click **Nonparametric Tests,** and then click **2 Independent Samples.** The "Two-Independent-Samples Tests" window in Figure 23.1 appears.

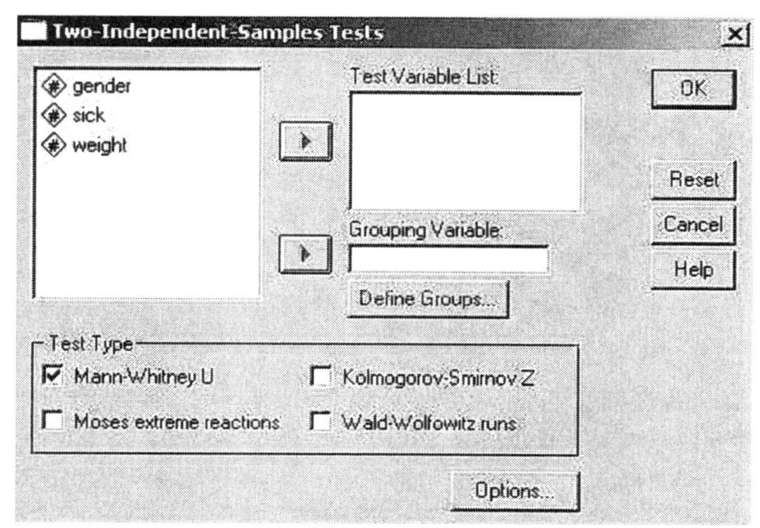

Figure 23.1

2. Click ***sick,*** then click ‣ to move ***sick*** into the "Test Variable List" box.
3. Click ***gender,*** then click ‣ to move ***gender*** into the "Grouping Variable" box.
4. Click the **Define Groups** button.
5. Type ***1*** in the "Group 1" box, press the **Tab** key, and type ***2*** in the "Group 2" box.
6. Click **Continue.**
7. The default test is the Mann-Whitney *U* test (as indicated by the ✓ in front of "Mann-Whitney U" in the "Test Type" area).
8. Click **OK.**

The resulting output is displayed in Tables 23.2 and 23.3.

Ranks

	GENDER	N	Mean Rank	Sum of Ranks
SICK	Female	196	163.25	31996.50
	Male	107	131.40	14059.50
	Total	303		

Table 23.2

Test Statistics[a]

	SICK
Mann-Whitney U	8281.500
Wilcoxon W	14059.500
Z	-3.334
Asymp. Sig. (2-tailed)	.001

a. Grouping Variable: GENDER

Table 23.3

As can be seen in Table 23.2, the rank sum for men (using average ranks for ties) is $W = 14{,}059.5$. Because the sample size is large, the z distribution should yield a reasonable approximation of the P-value. As shown in Table 23.3, the standardized value is $z = -3.334$ with a two-sided P-value = 0.001. This small P-value lends strong evidence that women are more concerned than men about the safety of food served at fairs.

Wilcoxon Signed Rank Test

This section will introduce the **Wilcoxon signed rank test,** the nonparametric counterpart of a paired-samples t test. It is used in situations in which there are repeated measures (the same group is assessed on the same measure on two occasions) or matched subjects (pairs of individuals are each assessed once on a measure). It is applied to situations in which the assumptions underlying the parametric t test have been violated or questionably met. The focus of this test is on medians rather than means.

The golf scores of 12 members of a college women's golf team in two rounds of tournament play are shown in Table 23.4. A golf score is the number of strokes required to complete the course; therefore, low scores are better.

Player	1	2	3	4	5	6	7	8	9	10	11	12
Round 2	94	85	89	89	81	76	107	89	87	91	88	80
Round 1	89	90	87	95	86	81	102	105	83	88	91	79
Difference	5	-5	2	-6	-5	-5	5	-16	4	3	-3	1

Table 23.4

The data were entered into SPSS for Windows using two columns and the variable names ***round_1*** (declared numeric 8.0) and ***round_2*** (declared numeric 8.0). The variables ***round_1*** and ***round_2*** were entered into the first and second columns, respectively.

Because this is a matched pairs design, inference is based on the differences between pairs. Negative differences indicate better (lower) scores on the second round. We see that 6 of the 12 golfers improved their scores. We would like to test the hypotheses that in a large population of collegiate women golfers:

H_0: scores have the same distribution in Rounds 1 and 2
H_a: scores are systematically lower or higher in Round 2.

The assessment of whether the assumption of normality has been met is based on the difference in golf scores. A small sample makes it difficult to assess normality adequately, but the normal quantile plot of the differences in Figure 12.2 shows some irregularity and a low outlier. Therefore, you should use the Wilcoxon signed rank test, which does not require normality.

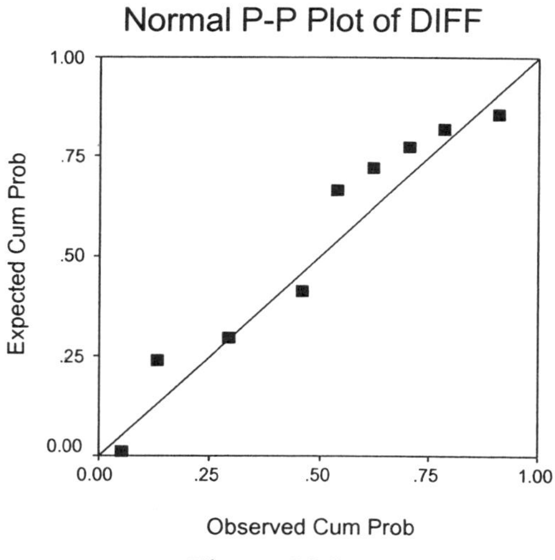

Figure 23.2

To conduct a Wilcoxon signed rank test, follow these steps:

1. Click **Analyze,** click **Nonparametric Tests,** and then click **2 Related Samples.** The "Two-Related-Samples Tests" window shown in Figure 23.3 appears.
2. Click ***round_1*** and it appears after "Variable 1" in the "Current Selections" box.
3. Click ***round_2*** and it appears after "Variable 2" in the "Current Selections" box.
4. Click ‣ to move the variables into the "Test Pair(s) List" box (it will read "***round_1 – round_2***").
5. The default test is the Wilcoxon signed rank test (as indicated by the ✓ in front of Wilcoxon in the "Test Type" box).
6. Click **OK.**

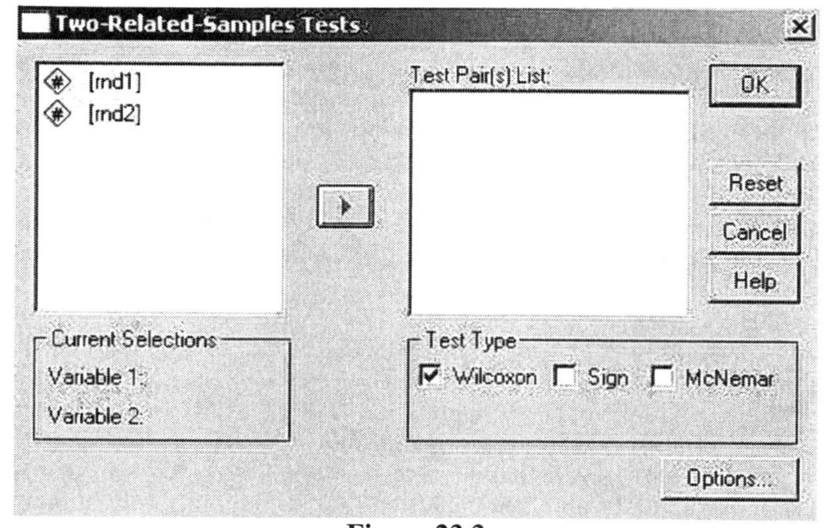

Figure 23.3

The resulting SPSS for Windows output is displayed in Tables 23.5 and 23.6.

Ranks

		N	Mean Rank	Sum of Ranks
-	Negative Ranks	6[a]	8.42	50.50
	Positive Ranks	6[b]	4.58	27.50
	Ties	0[c]		
	Total	12		

a. <

b. >

c. =

Table 23.5

Test Statistics[b]

	-
Z	-.910[a]
Asymp. Sig. (2-tailed)	.363

a. Based on positive ranks.

b. Wilcoxon Signed Ranks Test

Table 23.6

First, notice that the difference in Table 23.5 reads "ROUND_2 – ROUND_1", which is the reverse of the difference variable that appeared to be created in Step 4 (***round_1*** – ***round_2***). For this test, SPSS for Windows always creates a difference score between the two named variables based on the order in which the variables are entered in the data set. The variable that appears *first* in the data set is always subtracted from the variable that appears *later* in the data set. For this problem, the variables were entered in the order of ***round_1*** and then ***round_2.*** Thus, SPSS for Windows creates a difference score of ***round_2*** minus ***round_1,*** despite the order in which the variables appear in the "Test(s) Pairs List" box in the "Two-Related-Samples Tests" window. The same conclusion will be reached regardless of the order in which subtraction was done because the two-tailed *P*-value will be the same whether the difference is ***round_2*** – ***round_1*** or ***round_1*** – ***round_2.*** However, caution must be used in performing a directional test.

As shown in Table 23.5, the sum of the Negative Ranks (ROUND_2 < ROUND_1) is 50.5 and the sum of the Positive Ranks (ROUND_2 > ROUND_1) is 27.5. The value of 0 for Ties means that there were no pairs of scores in which the values were the same (e.g., ROUND_2 = ROUND_1 [it does not mean that there were no ties among the ranks]). The Wilcoxon signed rank statistic is the sum of the positive differences. The value is $W^+ = 27.5$. In Table 23.6, a z value of – 0.91 is reported that is based on the standardized sum of the positive ranks and is not adjusted for the continuity correction. The corresponding *P*-value is given as 0.363. These data give no evidence for a systematic change in scores between rounds.

Kruskal-Wallis Test

The **Kruskal-Wallis test** is the nonparametric counterpart of the parametric one-way analysis of variance. It is applied to situations in which the normality assumption underlying the parametric one-way ANOVA has been violated or questionably met. The focus of this test is on medians rather than means.

Lamb's-quarter is a common weed that interferes with the growth of corn. A researcher planted corn at the same rate in 16 small plots of ground, then randomly assigned plots to four groups. He weeded the plots by hand to allow a fixed number of Lamb's-quarters to grow in each meter of a corn row. These numbers were 0, 1, 3, and 9 in the four groups of plots. No other weeds were allowed to grow, and all plots received identical treatment except for the weeds. The yields of corn (bushels per acre) in each of the plots are shown in Table 23.7. The summary statistics for the data are shown in Table 23.8.

Weeds per meter	Yield (bushels/acre)			
0	166.7	172.2	165.0	176.9
1	166.2	157.3	166.7	161.1
3	158.6	176.4	153.1	156.0
9	162.8	142.4	162.7	162.4

Table 23.7 Yields of Corn

Weeds per meter	n	Mean	Std. Dev.
0	4	170.200	5.422
1	4	162.825	4.469
3	4	161.025	10.493
9	4	157.575	10.118

Table 23.8 Descriptive Statistics for Yield

The sample standard deviations do not satisfy the rule of thumb from BPS for use of ANOVA that the largest standard deviation should not exceed twice the smallest. Moreover, we see that outliers are present in the yields for 3 and 9 weeds per meter. These are the correct yields for their plots, so we have no justification for removing them. We may want to use a nonparametric test.

The hypotheses are:

H_0: yields have the same distribution in all groups
H_a: yields are systematically higher in some groups than in others.

The data were entered in two columns using the variables ***weeds*** (declared numeric 8.0) and ***yield*** (declared numeric 8.1). It is important to note that the grouping variable (in this case ***weeds***) must always be a numeric variable. To conduct a Kruskal-Wallis H test, follow these steps.

1. Click **Analyze,** click **Nonparametric Tests,** and then click **K Independent Samples.** The window shown in Figure 23.4 will appear.

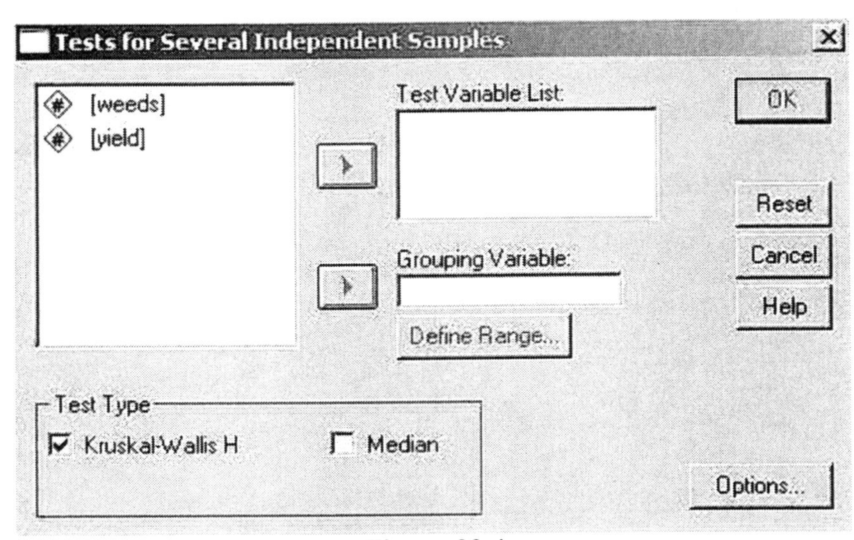

Figure 23.4

2. Click ***yield,*** then click ▸ to move ***yield*** into the "Test Variable List" box.
3. Click ***weeds,*** then click ▸ to move ***weeds*** into the "Grouping Variable" box.
4. Click **Define Range.**
5. Because 0 to 9 is the range for this example, type ***0*** in the "Minimum" box, press the **Tab** key, and type ***9*** in the "Maximum" box.
6. Click **Continue.**
7. The default test is the Kruskal-Wallis H Test (as indicated by the ✓ in front of Kruskal-Wallis H in the "Test Type" box).
8. Click **OK.**

Examination of Table 23.8 suggests that an increase in weeds results in decreased yield. As can be seen in Table 23.9, the mean rank for the group with 0 weeds per meter was 13.13, the mean rank for the group with 1 weed per meter was 8.38, the mean rank for the group with 3 weeds per meter was 6.25, and the mean rank for the group with 9 weeds per meter was 6.25. SPSS for Windows uses the Chi-square approximation to obtain a P-value = 0.134, as shown in Table 23.10. This small experiment suggests that more weeds decrease yield but does not provide convincing evidence that weeds have an effect.

Ranks

	N	Mean Rank
0	4	13.13
1	4	8.38
3	4	6.25
9	4	6.25
Total	16	

Table 23.9

Test Statistics[a,b]

	YIELD
Chi-Square	5.573
df	3
Asymp. Sig.	.134

a. Kruskal Wallis Test

b. Grouping Variable: WEEDS

Table 23.10

Chapter 24. Statistical Process Control

Topics covered in this chapter:

- **Pareto Charts**
- **Control Charts for Sample Means and Standard Deviations**
- **Control Charts for Sample Proportions**

Pareto Charts

Pareto charts are bar graphs with the bars ordered by height. They are often used to isolate the "vital few" categories on which we should focus our attention. Exercise 24.4 of BPS describes the following example.

A large medical center financially pressed by restrictions on reimbursement by insurers and the government, looked at losses broken down by diagnosis. Government standards place cases into diagnostic related groups (DRGs). For example, major joint replacements (mostly hip and knee) are DRG 209. The data can be found in the SPSS folder of your CD-ROM labeled ***ex24-04.por.*** Table 24.1 shows the data for Exercise 24.4 and list the 9 DRGs with the most losses along with the percent of losses. Because the percents are given, a Pareto chart can be constructed.

Drug	104	107	109	116	148	209	403	430	462
Percent Loss	5.2	10.1	7.7	13.7	6.8	15.2	5.6	6.8	9.4

Table 24.1

To construct a Pareto chart, follow these steps:

1. Click **Graphs,** then click **Pareto.** The "Pareto Charts" window in Figure 24.1 appears.

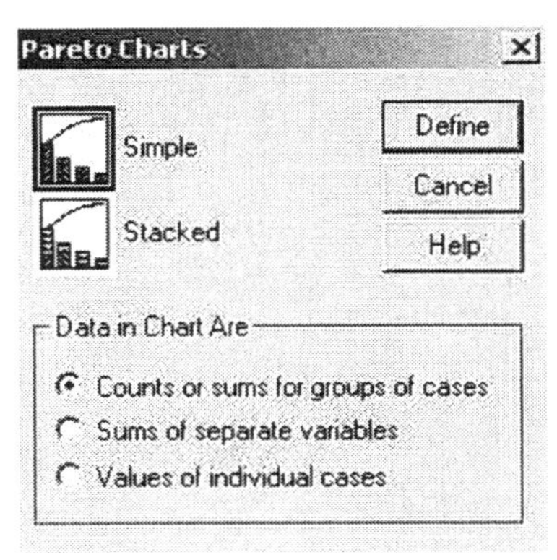

Figure 24.1

2. In the "Data in Chart Are" box, click **Values of Individual Cases.**
3. Click **Define**. The "Define Simple Pareto: Values of Individual Cases" window in Figure 24.2 appears.
4. Click ***drg,*** then click ▸ to move ***drg*** into the "Variable" box.
5. Click ***percentloss,*** then click ▸ to move ***percentloss*** into the "Values" box.

6. Click **OK.**

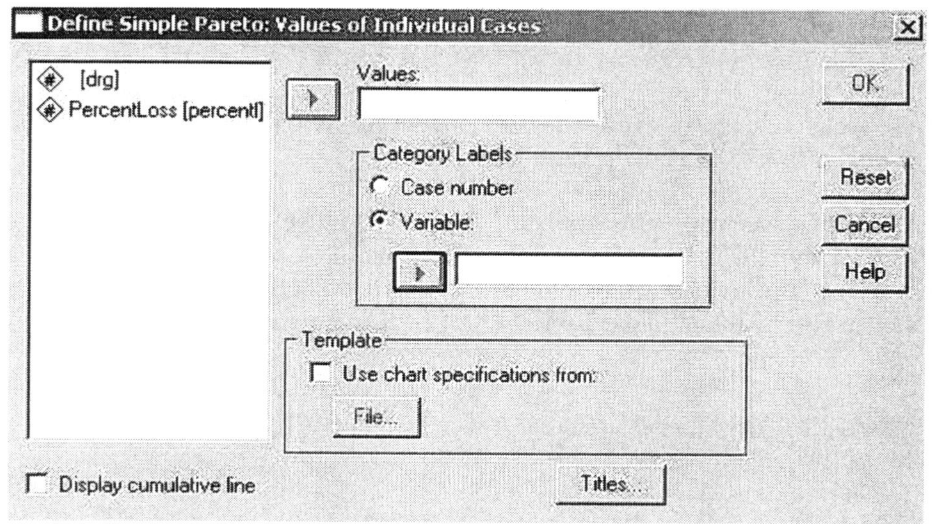

Figure 24.2

Figure 24.3 is the resulting SPSS for Windows output.

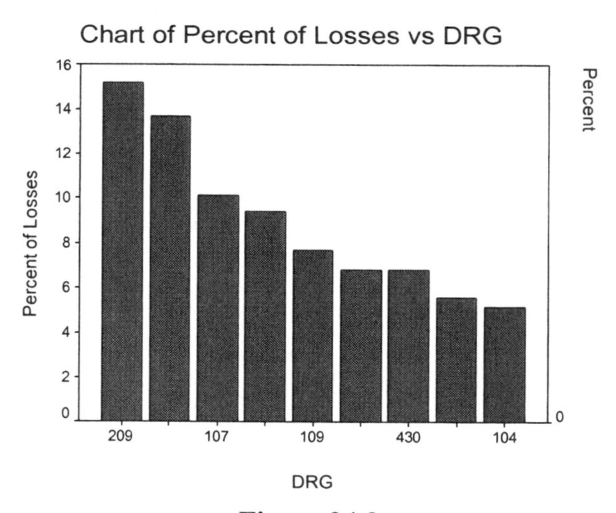

Figure 24.3

Control Charts for Sample Means and Standard Deviations

Control charts are statistical tools that monitor a process and alert us when the process has been disturbed so that it is now **out of control.** This is a signal to find and correct the cause of the disturbance. We first met control charts for sample means in Chapter 10 of this manual.

We will revisit the example concerning a manufacturer of computer monitors who must control the tension on the mesh of fine vertical wires that lies behind the surface of the viewing screen. Too much tension will tear the mesh, and too little will allow wrinkles. Tension is measured by an electrical device with output readings in millivolts (mV). The manufacturing process has been stable with mean tension μ = 275 mV and process standard deviation σ = 43 mV.

The mean 275 mV and the common cause variation measured by the standard deviation 43 mV describe the stable state of the process. If these values are not satisfactory — for example, if there is too much variation among the monitors —the manufacturer must make some fundamental change in the process. This might involve buying new equipment or changing the alloy used in the wires of the mesh. In fact, the common cause variation in mesh tension does not affect the performance of the monitors. We want to watch the process and maintain its current condition.

The operator makes four measurements every hour. Table 24.2 in BPS gives the last 20 hours. The first row of observations is from the first hour, the next row is from the second hour, and so on. There are a total of 80 observations. The table also gives the mean (x-bar) and standard deviations (s) for each sample. The data can also be found in the Table 24.4 of BPS and in the SPSS folder of your CD-ROM labeled **ta24.01.por.**

To produce a control chart using SPSS, follow these steps:

1. Click **Graphs,** then click **Control.** The "Control Charts" window in Figure 24.4 appears.

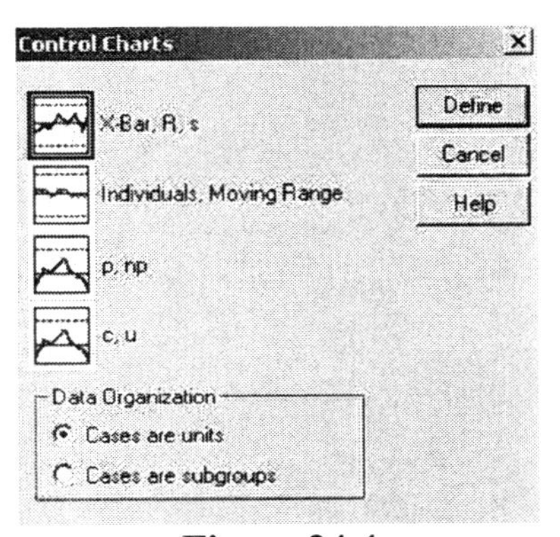

Figure 24.4

2. By default, the X-Bar, R, *s* type of chart is selected.
3. By default, in the "Data Organization" box, **Cases are Subgroups** is selected.
4. Click **Define.** The "X-Bar, R, *s*: Cases are Subgroups" window in Figure 24.5 on the following page appears.
5. In the "Charts" box, click on **X-Bar and standard deviation.**
6. Click ***sample***, then click ▸ to move ***sample*** into the "Subgroups Labeled by:" box.
7. Highlight **tension1, tension2, tension3,** and **tension4,** then click ▸ to move them all into the "Samples:" box.
8. Click **OK.**

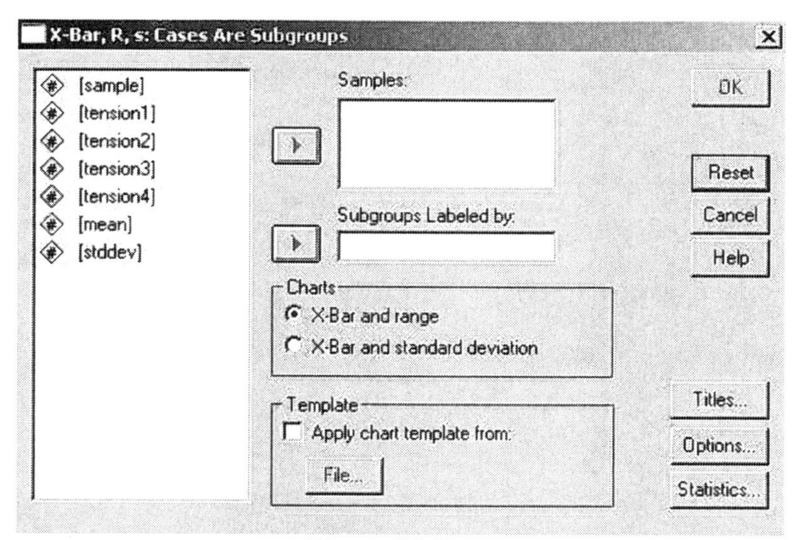

Figure 24.5

Figure 24.6 is the resulting SPSS for Windows output.

In this x-bar chart, no points lie outside the control limits. In practice, we must monitor both the process center, using an x-bar chart, and the process spread, using a control chart for the sample standard deviation *s*.

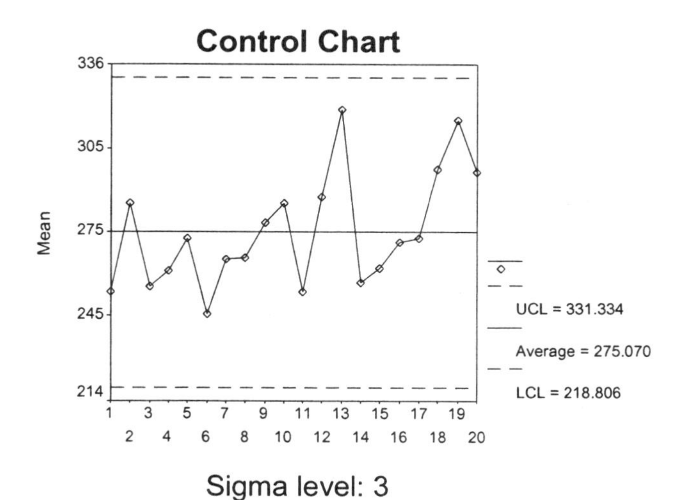

Figure 24.6

In practice, we must control the center of a process and its variability. This is commonly done with and *s* chart, a chart of standard deviation against time. Figure 24.7 on the next page shows the resulting SPSS for Windows output.

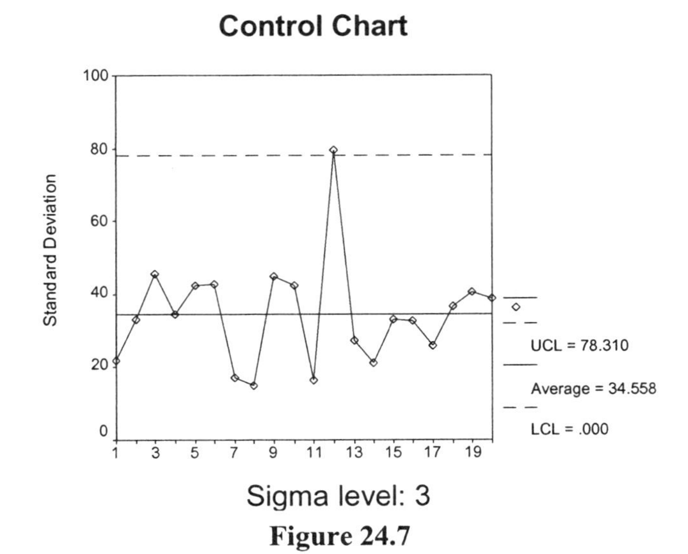

Figure 24.7

Control Charts for Sample Proportions

Example 24.13 of BPS uses *p* charts for discussing manufacturing and school absenteeism. Table 24.8 of BPS and **ta24.08.por** in the SPSS folder of your CD-ROM contain data on production workers and record the number and proportion absent from work each day during a period of four weeks.

To use SPSS to produce control charts for proportions, follow these steps:

1. Click **Graphs,** then click **Control.** The "Control Charts" window shown above in Figure 24.4 as shown above appears.
2. By default, the X-Bar, R, *s* type of chart is selected. To obtain a *p* chart, select the icon p, np type of chart.
3. By default, in the "Data Organization" box, **Cases are Subgroups** is selected.
4. Click **Define.** The "p, np: Cases are Subgroups" window in Figure 24.8 appears.

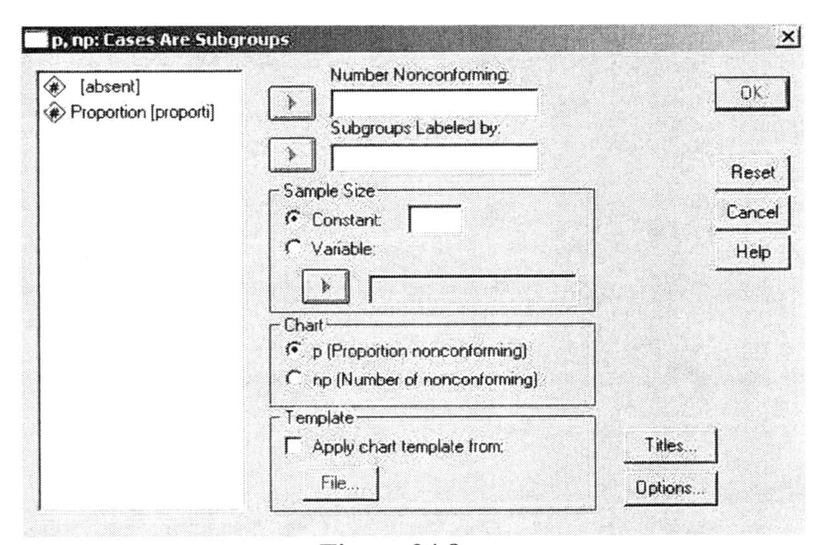

Figure 24.8

5. Click ***absent,*** then click ▸ to move ***absent*** into the "Number Nonconforming:" box.
6. Click ***proportion,*** then click ▸ to move ***proportion*** into the "Subgroups Labeled by:" box.
7. By default, in the "Sample Size" box, **Constant** is selected. In the empty box beside **Constant,** enter the sample size. In this example, the sample size is 987.
8. Click **OK.**

Figure 24.9 is the resulting SPSS for Windows output.

The *p* chart shows a clear downward trend in the daily proportion of workers who are absent. It appears that actions were taken to reduce the absenteeism rate. The last two weeks' data suggest that the rate has stabilized.

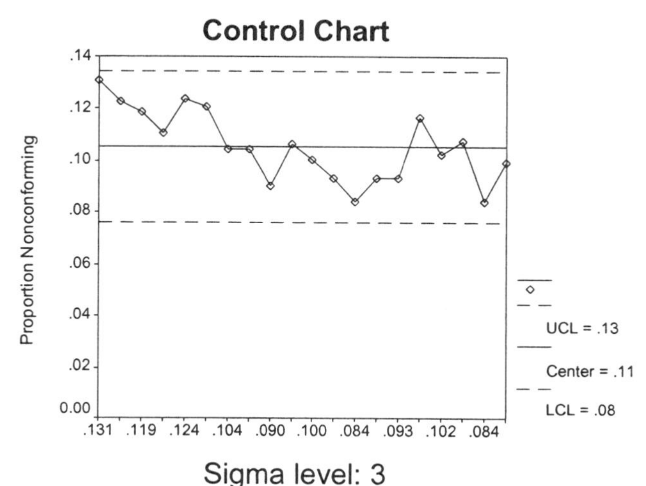

Figure 24.9

Index

A

B

C

D

E

F

G

H

I

L

M

N

O